COLLINS BRITISH BIRD IDENTIFIER

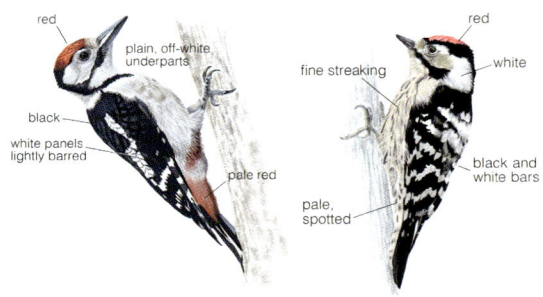

Paul Stancliffe & Jeff Baker

Dedicated by Paul Stancliffe to Abigail

William Collins
An imprint of HarperCollinsPublishers
1 London Bridge Street
London SE1 9GF

WilliamCollinsBooks.com

HarperCollinsPublishers
1st Floor, Watermarque Building, Ringsend Road
Dublin 4, Ireland

First published in Great Britain in 2026 by William Collins

1

Copyright © Paul Stancliffe & Jeff Baker 2026

The authors assert their moral right to be identified as the author of this work in accordance with the Copyright, Designs and Patents Act 1988

A catalogue record for this book is available from the British Library

ISBN 978-0-00-850807-4

All rights reserved. No part of this publication may be reproduced, stored in a retrieval system, or transmitted, in any form or by any means, electronic, mechanical, photocopying, recording or otherwise, without the prior permission of the publishers.
Without limiting the author's and publisher's exclusive rights, any unauthorised use of this publication to train generative artificial intelligence (AI) technologies is expressly prohibited. HarperCollins also exercise their rights under Article 4(3) of the Digital Single Market Directive 2019/790 and expressly reserve this publication from the text and data mining exception. This book is sold subject to the condition that it shall not, by way of trade or otherwise, be lent, re-sold, hired out or otherwise circulated without the publisher's prior consent in any form of binding or cover other than that in which it is published and without a similar condition including this condition being imposed on the subsequent purchaser.

Printed and bound in Bosnia & Herzegovina by GPS Group

Paul Stancliffe Acknowledgments

This book is the result of decades of notetaking, writing down the salient features needed to help me identify birds with confidence. As a novice birdwatcher these notes were essential, over fifty years later I still use them as an 'aide memoir'. I have had the good fortune to have birded with many very talented birders and owe a debt of thanks to them. My mentor, Dave Smith, has shared his extensive identification knowledge over the years. My long-term birding buddy, Andy Mason has shared many great birds with me and debated many a tricky identification – thank you both. Andy Clements gave me the confidence to share any knowledge I might have with others and of course shared some great birding moments too. Jeff Baker's artwork has brought this book to life, thank you for sticking with it through thick and thin and producing some of the best artwork I have seen. I owe the biggest thank you to my family. My wife Abigail has been by my side for almost forty years and has been an integral part in many birding adventures, more often than not being responsible for making them happen. She has also been a driving force in making this book happen too. My daughters, Lily and Hazel, have a much larger bird list than they are probably aware of, having tolerated hours in the field with me. I love all three of them immensely and thank them for their never-ending patience and their inspiration. Of course, there are others too but too many to list here, but I hope you know who you are. Of course, there are others, too many to list here but I hope you know who you are.

Jeff Baker Acknowledgements

This work is a result of accumulated knowledge over many decades, both in the field and hundreds of visits to the Natural History Museum at Tring (formerly the Sub-department of Ornithology of the British Museum). The staff at the museum were always very welcoming and helpful, in particular Peter Colston, Ian Bishop, Graham Cowles, and more recently, Robert Prys Jones. Like so many people, a schoolteacher sowed the seeds that changed my life and nurtured my interest in birds: Peter Walton taught art and was himself a very talented bird artist and field birder. He will be forever in my thoughts. Moss Taylor and Sean McMinn, have both been good friends to me over many years and their field skills are second to none. They commented extensively on the illustrations, and I thank them both for their invaluable input. As an illustrator accuracy in depicting the finer details of a bird is essential, especially so when the identification between two very similar species may come down to one or two subtle features. A resource to help identify, and then draw some of these has been made a good deal easier today by reference to the vast online collections of bird photographs. To the thousands of very dedicated bird photographers out there I thank you. As an artist and author of other books too, I have lost much time with my family; they have paid a heavy price and no amount of gratitude will ever make up for that! Nonetheless, I extend a heartfelt thank you to Fran, Jenna, Nyla and Kobe. Finally of course thanks go to Myles Archibald and his team at Collins for encouragement and support throughout the project.

Introduction

This bird identification guide is like no other. You won't find descriptions of preferred habitats, information on feeding preferences or geographical range or lengthy discussions on taxonomy and nomenclature. There are excellent treatises on these already. What you will find is a clean and innovative way of helping to identify any bird on the British List.

During my 50 years of watching and keeping notes on the birds I have seen around the world, but more importantly here in Britain, I have found that, more often than not, it pays to focus on one or two features initially (with some views that is often all there is time for) and then expand out as and when, and if, better views allow.

The first view of a warbler can give a clue as to what family it belongs to, brown (*Acrocephalus*), green (*Phylloscopus*) and grey (*Sylvia*). Better views reveal a plain brown warbler with paler underparts – so, what to look for? The length of the wing is important – this bird has long wings with pale tips to the flight feathers (the primaries). The bill and face pattern? It shows a fairly stout bill and an obvious pale supercilium (stripe over the eye) that bulges between the bill and eye (the lores) and has plain olive grey/brown upperparts (including the rump). With all these features combined we can fairly confidently identify this as a Marsh Warbler.

Every species currently on the British List is included. Generally, subspecies are not listed unless they occur commonly in the UK and, for vagrant species, their 'usual' observed plumage is illustrated.

Species that share a plate are illustrated at a scale in proportion to each other to give an idea of size comparison between those species depicted. The size proportion is only relevant to each plate.

The descriptions of colour are nuanced as you might expect. Red-brown is a shade of brown, whilst red/brown describes a feature that is variable in colour, i.e. either red or brown.

The call of a bird can often be key to its identification and if this is one of the features a phonetic description is written by 'call'. You are encouraged to find audio resources to complement your field skills in these cases.

Using this book

This book gives you the five features you need to identify every bird on the British List. Two or three features might be enough for some birds, but with five (not always possible to get 'in the field' on every bird) identification can be reached without doubt. The type of features differ with the different birds but they are all contained within this book. So let's get started with a few notes on how to get the most out of each plate of illustrations.

We have loosely followed the British Ornithologist's Union order but have moved some species closer to others where we think it will help in separating and identifying similar birds. We have also done this with a small number of groups. We are aware of the current taxonomical state of flux and that the order birds are listed in and some of the common names in use will change. These changes should have little impact on the usefulness of this book in helping to identify birds.

The aim is to keep, as much as possible, the amount of text in the book to a minimum but we have included additional information where we think it is useful in at the front of the book. We have not included information on all species, or indeed all groups of species – this is intentional as much of what we could write already exists in other books and, some species speak for themselves.

Male/Female: when the sexes are alike a single illustration is used with no male or female symbol.

Young birds: juveniles are birds that have not begun their post-juvenile moult and so have no older feathers. Immatures are birds that have begun their post-juvenile moult and have some adult-like feathers. A fledged young bird leaves the nest with very-quickly grown (c.14 days) 'not the best' quality feathers. Some birds will immediately begin a post-juvenile moult replacing feathers with ones that will take them through their first winter (see also Abbreviations and Symbols).

Seasons: from time-to-time birds arrive in the UK out of season, coloured dots show the seasons in which a species occurs regularly.
Key:
Green - spring
Yellow - summer
Orange - autumn
Blue – winter

For example:
- 🟢 🟡 🔵 🔷 Seen all year
- 🟠 Autumn only

Status: We have coloured the species names to give a general reflection of the status of each.

Red = rare
grey = scarce
black = occurs commonly

For example:
Crested Lark = rare
Woodlark = scarce
Skylark = common

Illustration labels: The body of the bird (the bit that the head, wings, tail and legs are attached to) can be split into upperparts and underparts. Where these differ in colour they are labelled separately, where the body of a bird is fairly uniform the term 'body' is used.
Where wing length is important there are diagrams added to show the length of the exposed primaries in relation to the exposed tertials - the length of the primaries and tertials that show on a bird's closed wing (NOT including the shoulder, just the primaries and tertials) represents 100%.

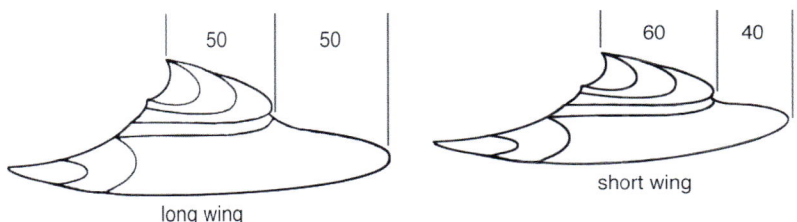

If the figure shown is 50% then the length of the primaries is equal to the length of the tertials. If the figure is 60% then the primaries appear longer than the tertials (40%). If the figure is 30% then the primaries appear shorter than the tertials (70%). The long or short wings labels help to discern between the different species.
If the primary percentage is 30%, these are 30% of the length of both of these feather tracts. Short primaries make the whole wing short,

whilst the opposite is true of long primaries.

Where wings are illustrated showing the wing plumage, the upper wing is the default and the left wing is always shown. Where the underwing is illustrated this is labelled as such.

Other plates show illustrations of leg length comparisons (compared to tail in flight etc etc), or bill length where these features are one of the five needed. Where tail pattern is important we have illustrated either the left hand half of the tail where the outer tail pattern is needed, or the full tail where the pattern spreads across the whole tail.

Notes on Groups

Grouse – p. 29
There are four species of grouse on the British List, namely **Capercaillie**, **Black Grouse**, **Ptarmigan**, and **Red Grouse**. They can all be difficult to see, with views of the head as they move through dense vegetation often the only sign of their presence. However, they will occasionally take flight and if you concentrate on the wings and the colour of these – all-brown/all-white – and the presence or absence of any wingbars - this will help you to separate them. In the winter both sexes of **Ptarmigan** are all-white.

Partridges – p. 30
Two species of partridge occur in Britain, **Red-legged** and **Grey**, though **Quail** and **Corncrake** are included as they are easily confused with the two partridges. The female **Grey Partridge** is slightly duller overall than the male. When meeting with a partridge species concentrate on the belly and flank patterning. Both **Quail** and **Corncrake** are very secretive and are therefore more likely to be heard than seen. Female **Quail** differs from the male by having a have a pale throat.

Pheasants – p. 31
There are three species of pheasant on the British List. Introduced for shooting and their ornamental plumage, all three have bred in the wild in the past. There is now some doubt that **Lady Amherst's** still breeds in the wild and whether the **Golden Pheasant** population continues to be self-sustaining. Escapes from captivity of these two species do occur, though rarely, and as such individual birds may be seen anywhere, and at any time of year.

Geese – p. 32
Geese that occur in Britain fall into two groups, *Branta* (geese with black plumage that include, **Brent, Red-breasted, Barnacle, Canada** and **Cackling**) and *Anser* (grey geese, **Greylag, Taiga** and **Tundra Bean, Pink-footed, White-fronted** and **Lesser White-fronted**). *Branta* geese have dark legs, black bills and obvious white markings around the head and neck. *Anser* geese have orange or pink legs and the bill can range from completely orange/pink, to some with small amounts of orange/pink. In *Branta* geese the amount of white in the head and neck, and the position of the white, will help to quickly narrow down

to species. In *Anser* geese, bill size and shape and the amount of orange/pink will do the same. Two subspecies of **Brent Goose** occur in Britain: *hrota* (light-bellied) occur widely in western Britain and the nominate *bernicla* (dark-bellied) occurs mainly in eastern Britain. *nigricans* (Black Brant) is a vagrant from North America. The dark morph **Snow Goose**, also known as blue morph, is very rare to Britain, most of those found being white birds. Immature birds (white morph) are duskier than the adult birds and have a grey bill. The juveniles of both **White-fronted** and **Lesser White-fronted Goose** lack the white forehead and have duller bills than the adults.

Swans - p. 36
Three species of swan are found in Britain, namely **Mute, Whooper** and **Bewick's**. **Mute Swans** are resident, the male on average, has a larger black knob on the bill than the female. **Whooper** and **Bewick's** are winter visitors, although a few **Whoopers** can be found during the summer in the north. All three can be separated by the amount of colour, and the pattern of the bill. **Whooper** and **Bewick's** swans regularly call in flight and with practice this can be used to separate them. **Bewick's** call is high-pitched and yodelling, whilst the call of **Whooper** is a deeper bugling sound.

Dabbling ducks – p. 37
Shelduck juvenile plumage is a subdued grey and white version of the adult, with a pink bill and greyish legs. Male and female **Shelduck** differ in the size of the red knob on the bill, large in the male and much reduced in the female.

When encountering a species of 'freshwater' duck, it can be helpful to determine whether it is a dabbling or a diving duck. Dabbling ducks feed on the surface and upend whilst diving ducks 'dive' to feed and rarely upend. Dabbling ducks 'spring' from the water without any need for a run-up, whilst diving ducks patter across the surface on take-off. In the female dabbling ducks the colour of the speculum, the panel of coloured feathers in the inner wing, is an important aid to identification. However, in the summer months dabbling ducks undertake a moult, dropping all their flight feathers for a time and become flightless, consequently the speculum is no longer present. The plumage attained during the summer moult, (where both sexes become female-like) is referred to as 'eclipse' plumage. Male and female dabbling ducks are difficult to identify in 'eclipse' plumage. The females of **Teal** and

Green-winged Teal are almost indistinguishable from each other in any plumage. This is also the case for **Wigeon** and **American Wigeon**.

Diving ducks – p. 44
The diving ducks include the pochards and the black-and-white ducks. To help separate the pochards, the colour and any patterning on the bill is helpful. In the black-and-white diving ducks, the colour and extent of any wingbar can clinch an identification and, although this is difficult to see on a swimming bird, ducks will often wing stretch and show the complete open wing. It is worth noting that female **Tufted Duck** sometimes show a white face similar to **Scaup** and that male **Ruddy Duck** attains a black neck-ring for a very short period during the breeding season.

Seaducks – p. 47
Ten species of duck found in Britain spend most of their life in and around coastal waters and are known as 'sea ducks'. This group includes the **Eiders, Scoters** and **Long-tailed Duck**. Within the 'sea ducks' there are three species which frequent both the sea and freshwater environments, namely **Bufflehead, Goldeneye** and **Barrow's Goldeneye,** which are often referred to as 'bay ducks'. All of the 'sea ducks' undertake a complete moult and attain a female-like eclipse plumage during the summer months. In full breeding plumage male eiders are quite colourful and easily separated. However, the brown females can be more challenging and bill shape is key.
Of the six species of scoter on the British List, the males are largely black and the females dark brown with pale cheeks. Like the eiders, bill shape is key but close views are often needed to assess this. In the males any colour in the bill and the extent of that colour is also needed. In both sexes any white in the wing is also helpful.
Note, some female **Common Scoter** are inseparable from female **Black Scoter** though with good views of the back of the head some are possible. The Asiatic form of **White-winged Scoter, Stejneger's Scoter**, has been recorded in British and Irish waters and is now considered a full species. This bird has darker flanks and a distinctive protruding horn on the top of the bill.

Sawbills – p. 53
Sawbills is the collective name given to **Smew,** the **Mergansers** and

Goosander. These too undertake a complete moult and attain a female-like eclipse plumage. Head colour, and the pattern of this, is key to identifying these birds both in breeding and eclipse plumage.

Nightjars – p. 55
Of the four different species of nightjar that have occurred in Britain only one breeds here, the Nightjar, the others are very rare. **Common Nighthawk** breeds in North America, but there have been over 20 recorded in Britain.
Juvenile **Common Nighthawk** shows a white trailing edge to the wing. **Red-necked Nightjar**, which breeds in the Iberian Peninsula and North Africa, has only been recorded once in Britain. The rufous nape on female **Red-necked Nightjar** is generally duller than that of the male.
There have been two **Egyptian Nightjars**, one was shot in Nottinghamshire in 1883 and the other was seen alive at Portland, Dorset in 1984. When nightjar species are encountered the presence or absence of white in the wing, and its location on the wing, is key.

Swifts – p. 56
When viewing swifts pay attention to any white in the plumage around the belly and rump area. This is an indicator that you may be viewing one of the scarcer/rarer species and attention needs to be paid to the exact location of the white and its extent. Whilst seven species of swift have occurred in Britain, rather surprisingly, only the **Swift** breeds in Britain, whilst the status of the other swift species is either rare or scarce.

Bustards – p. 58
There are three species of bustard on the British List. **Great Bustard** once bred regularly in southern and eastern England but the species is now confined to the few birds that have been reintroduced to Salisbury Plain in Wiltshire. **Little Bustard** is a rare visitor. **Asian Houbara** is very rare indeed, only five individuals have ever been seen in Britain. When viewing bustards the amount of white in the wing is crucial to its identification.

Cuckoos – p. 59
Cuckoo, Great Spotted Cuckoo, Yellow-billed Cuckoo and **Black-billed Cuckoo** have all been recorded in Britain. **Cuckoo** is a summer

visitor, whilst **Great Spotted** is a rare visitor from southern Europe, **Yellow** and **Black-billed** are both rare vagrants from North America. The grey morph of female **Cuckoo** is similar to male **Cuckoo** but often shows some buff colour on the throat and upper breast. The adult female rufous morph of **Cuckoo** is similar to the juvenile but lacks the pale spots in the wing and nape and is a brighter rufous colouration above. Adult **Black-billed** and **Yellow-billed Cuckoos** are rarely seen on this side of the Atlantic. The illustrations relate to juvenile plumages, though the adult birds don't differ appreciably.

Pigeon – p. 60
Pigeons and Doves form a single group with the term 'dove' used as an alternative name for a pigeon. Therefore, all members of this group are, strictly speaking, pigeons. The whole group can be neatly separated into two; those that are blue-grey and those that are sandy-brown. Pointers for identification in the blue-grey group (**Rock Dove, Stock Dove** and **Woodpigeon**) are principally the number of wingbars and eye colour. In the sandy-brown group (**Collared Dove, Turtle Dove, Oriental Turtle Dove** and **Mourning Dove**) the length of the wings and the extent and colour of the tail corners is helpful.

True **Rock Dove** is a bird of sea cliffs and can now only be seen in the remotest parts of northern Britain. 'Feral Pigeons' are all descendants of the Rock Dove and these can be seen throughout Britain in a variety of plumages. In **Stock Dove** all but the youngest birds show a neck patch that is iridescent green, whilst the adult **Woodpigeon** additionally has a white neck patch. Adult **Collared Dove** have a black-and-white collar whilst young **Mourning Doves** are similar to the adults. All but the youngest **Turtle Doves** show black-and-white neck stripes, with young birds also showing pale fringes on the wings. Adult **Oriental Turtle Dove** lacks the pale fringes of the young birds on the upperparts.

Rails and crakes – p. 62
The only **Western Swamphen** that has been recorded in Britain to date was an adult, so this is the plumage that has been illustrated. A juvenile may occur in the future and in this plumage will look like a duller version of the adult.
Allen's and **American Purple Gallinule** have only occurred in immature plumage. As an adult bird **Allen's Gallinule** has greenish upperparts and purple/blue underparts with a red bill and a small, round

white shield and red legs. Whilst adult **American Purple Gallinule** is similar to the immature bird except for the yellow tip to its red bill, and its yellow legs.

American Coot is shown as a 1st-winter bird as this is the most common plumage recorded in Britain. Adult plumage is a deeper black with a green-blue gloss. The bill and shield are purer white with a more pronounced black bar across the bill tip. The eye is flame red.

Cranes – p. 65
Immature **Sandhill Cranes** have similar plumage to the adults once the post-juvenile moult is completed. The **Sandhill Cranes** seen in Britain have shown a varying amount of rusty feathering in the wings. Immature **Common Crane** have a rusty head and neck and lack the head pattern of the adult.

Grebes – p. 65
Six species of grebe occur in Britain, five can be seen regularly, **Little Grebe, Red-necked Grebe, Great Crested Grebe, Slavonian Grebe** and **Black-necked Grebe**. **Pied-billed Grebe** is a rare vagrant from North America. In summer plumage all six are easily separated and with good views winter plumaged birds should present few problems. In the winter bill length and shape is helpful, along with any duskiness and the extent of it in the neck.

Oystercatcher to Sand Plovers
The young of most of the following species are only in their juvenile plumage for a matter of weeks, so the adults are illustrated in the main. Once their post-juvenile moult is completed they appear adult-like.

Oystercatcher, allies and plovers – p. 67
Oystercatcher - juvenile plumage duller and browner than adult; **Black-winged Stilt** - juvenile plumage has pale fringes to the dark upperparts; **Avocet** - juvenile plumage duller and browner. **Lapwing** - juvenile/adult winter birds have pale fringes to the upperpart feathering and a slightly shorter crest than adult summer lapwings. **Sociable Lapwing** - adult winter and 1st winter plumage is similar, having pale fringes to the back and wing feathers. Summer plumage birds show rather plain upperparts and a dark belly patch. **Golden Plover** - whilst juvenile and winter plumages are very similar, adult summer females have smaller areas of smudgy black on their underparts. **Pacific**

Golden Plover - moulting summer to winter adults can show blotchy underparts though the primary/tertial ratio is unchanged whilst adult winter and 1st-winter Pacific Golden Plovers are very similar to **American Golden Plover. Pacific American Golden Plover** adults moulting from summer to winter plumage can have blotchy underparts but always maintain the primary/tertial ratio. 1st-winter **Grey Plover** can look surprisingly yellow (therefore resembling young Golden Plover) but in flight will reveal black 'armpits' and a white upper tail. **Ringed Plover** - all ages are similar though juveniles show pale fringes on their upperparts. **Semipalmated Plover** - for this species the call is helpfully diagnostic, a disyllabic 'cheewi'. **Killdeer** - juvenile/1st-winter plumage is similar to the adult but with some buff between the black breast-bands. **Kentish Plover** - juvenile and winter plumages similar to the adult female.

Sand Plovers – p. 71
Very rare in Britain and are notoriously difficult to identify. Recently split into two full species, **Siberian Sand Plover** *Charadrius mongolus* & **Tibetan Sand Plover** *Charadrius atrifrons* both are illustrated as **Lesser Sand Plover**. Close attention should be paid to the head pattern and bill proportions.

Dotterel – p. 72
Only the female and winter **Dotterel** plumages are illustrated. Immature plumage is similar to the adult winter except for the upperparts being fringed bright creamy-buff. Summer males are duller versions of summer females.

Curlews – p. 72
For **Whimbrel** and **Curlew,** the immature plumage is similar to adult plumage apart from pale fringes to the upperparts and more buff on the neck and breast. The **Upland Sandpiper** is shown in immature plumage as this is the plumage type most likely to be seen in Britain. The adult plumage is very similar to the juvenile but the pale fringes on the upperparts of the adult are less pronounced.

Godwits – p. 73
Adult summer and winter **Godwits** are illustrated. The immature birds are similar to the adult winter birds except for a peachy wash on the neck and breast and more patterned upperparts showing pale fringing.

Sandpipers and allies – p. 74
Stilt Sandpiper in winter plumage is a rather plain, grey version of the immature plumage. **Curlew Sandpiper** in winter plumage is a cleaner, greyer version of the immature plumage but with a more obvious split supercilium.
Adult male **Ruff** summer plumage can be highly variable and the leg colour in the females can be variable too, with some showing dark green legs. However, they always have a small looking head, including the male birds once they have lost their breeding 'ruffs'. The male **Ruff** is always much larger than the female. Reeve is the name given to female **Ruff**.

Stints – p. 76
Stints are the smallest waders, and some of the rarest, to occur in Britain. As a group they are also amongst the most difficult to identify and close and critical observation of the features is necessary. Particular attention should be paid to the leg colour, head pattern and the extent of any breast streaking.
Dunlin is the small wader to get to know, and know well, as a comparison species. As such it is considered to be the 'default' small wader. It is very common in Britain and can be seen at any time of the year in a variety of plumages. Three subspecies regularly occur (*schinzii, alpina* and *arctica*) and differ in size and bill length, with *schinzii* being the smallest and having the shortest bill and *alpina* the largest with the longest bill but be aware that there is overlap. **Sanderling** is the only species of small wader to lack a hind toe!

Dowitchers – p. 79
Two species of dowitcher have been seen in Britain, **Long-billed Dowitcher** and **Short-billed Dowitcher**. Both can be separated on plumage with care, in particular by the pattern on the tertial feathers, but the easiest way to separate them is by their calls.

Pratincoles – p. 85
Only adult plumage pratincoles are illustrated as immatures are very similar apart from some pale upperpart fringing. When identifying immature pratincoles the same principle applies to adult identification – the presence or absence of a white trailing edge to the secondaries and the colour of the underwing are critical.

Gulls – p. 86
Small gulls (Kittiwake to Ring-billed Gull) take three years to reach full adult plumage, which is acquired through a succession of annual moults. Only first year immature birds and adults are illustrated, as the plumages in between these are either very similar to the first year immatures or the adult. **Large gulls (Great Black-backed to Caspian Gull)** take five years to reach full adult plumage, with adult plumage acquired through annual moults. Not every stage of the five-year moult has been illustrated: the most useful set of plumages have been shown.

Terns – p. 93
The typical terns, **Common, Arctic** etc, achieve adult plumage after three annual moults. First year and adult plumages are illustrated. Second year plumage is rarely seen in Britain as most second year birds remain close to their winter quarters. The majority of **Sooty Terns** and **Bridled Terns** seen in Britain have been adults so this is the plumage shown here. Whilst not a tern, **Red-billed Tropicbird** is included on this plate as it is very tern-like in appearance and flight.

Marsh Terns – p. 97
Black, **White-winged** and **Whiskered Terns** (the three species of marsh tern seen in Britain) are illustrated in adult and immature plumage as these are the most useful. All three are passage migrants in Britain occurring in varying numbers depending on the weather – easterly winds during migration often drift these birds to Britain.

Gannets – p. 98
Gannets reach adult plumage over 5-6 years and with the appearance ranging from all dark to patchy black and white to, ultimately, all-white. In years 5 and 6 the only sign of immaturity might be a single black feather in an otherwise white tail. In the **Gannet** plate, the moult cycle has been compressed to best illustrate the most common plumage types seen. To date, only immatures of **Red-footed Booby** have been seen in Britain. There have been two British records, one 'brown' morph and a 'pale' looking bird, the illustration fits neatly between the two. All Brown Booby records have been of birds in immature plumage.

Skuas – p. 99

The **smaller skuas** (**Pomarine, Arctic** and **Long-tailed**) can show very variable plumage, some may have a completely dark head whilst others can be almost white. Some may show a prominent white rump whilst others are all dark. The plumages encountered most are depicted. The two larger skuas, Great and South Polar invariably differ in plumage tone, warmer brown in Great, colder and darker in South Polar.

Divers – p. 103

Identifying divers can be tricky but head and bill shape is extremely helpful in separating them. In flight, the amount of foot projecting beyond the tail can also be useful.
Pacific Diver can show a prominent dark 'chin-strap' between the underside of the throat and upper neck though this feature is not always present.

Frigatebirds – p. 104

When viewing frigatebirds first focus on the amount and distribution of any white in the plumage. We have shown the plumage types that have occurred in Britain to date.

Ibis – p. 110

Only adult **Glossy Ibis** is illustrated in this guide. Younger birds appear similar to the adults but are duller and spotted.

Cormorants – p. 110

The subspecies *C.c.sinensis* (**Continental Cormorant**) regularly occurs in Britain and is most predominant type of cormorant found on inland waters. It is very similar to **Great Comorant** *C.c.carbo* and can only be confidently separated by assessing the gular pouch angle which is illustrated. The gular pouch angle is more acute (around 60 degrees) in **Great Cormorant** and close to 90 degrees in **Continental Cormorant**. **Double-crested Cormorant** has only been found in Britain in immature plumage and this is the plumage illustrated. Adult plumage is very similar but shows brighter bare parts around the head.

Herons – p. 111

There are fourteen species of heron on the British List. Five of these are very rare: Great Blue, Chinese Pond, Least Bittern, Green Heron and Snowy Egret. The illustrations of these are of immature birds as those seen have all been in this plumage. Of the remaining nine heron species, all have bred or attempted to breed, except for Squacco Heron. All of the white herons and egrets show a variable number of head and body ornaments during the breeding season.

Birds of Prey (Vultures to Accipiters) — p. 115
Of the Vultures, only the **Egyptian Vulture** currently features on the British List. When considering rarer eagles only immature **Short-toed** and **Spotted Eagle** have been recorded, so these are the plumages shown. **Booted Eagle** is a very recent addition to the British List and is illustrated at the end of the book.
The plumages of **Honey Buzzard** and **Buzzard** range from very light to very dark, and everything in between, so the most likely plumages to be encountered are shown.
The size difference between male and female **Sparrowhawk** and **Goshawk** is greater than in any other bird of prey in Britain. Males are always much smaller than the females in these species. However, care is needed when using size alone as there is overlap with the largest and smallest males or females in both species.

Kites – p. 118
In most circumstances, the two kite species, **Red** and **Black Kite**, can be readily identified by the depth of the fork in the tail. When the tail is fanned the fork can look shallow, so assessing the fork depth on a closed tail is best.

Harriers – p. 119
Females and immature harriers, apart from **Marsh Harrier**, can be very difficult to identify, with wing structure and pattern giving helpful pointers. Is the hand (the outermost part of the wing) narrow or broad? How many 'fingers' are present? Take note of any barring in the wing and its extent and location. Assessing the head pattern of female and immature harriers is also useful. Look for the amount of white in the face and the colour of the nape. With good views, the males should present few problems.

Owls – p. 120
When looking at 'small owls' (**Hawk**, **Tengmalm's**, **Little** and **Scops Owl**) face pattern is key to identification and when seen well is 'clinch-

ing'. Juvenile plumages are not illustrated for this group as these moult stages are only seen for a short period of time. Face pattern is also important when identifying 'large owls' (**Short** and **Long-eared; Tawny**, **Barn** and **Snowy Owl**), as is eye colour. Juvenile plumages are not illustrated as this plumage stage is so short. **Eagle Owl** is not included here as there is reasonable doubt about its occurrence in the wild.

Hoopoe to Bee-eaters – p. 122
Hoopoe, Roller, Kingfisher and Belted Kingfisher pose few challenges but **Blue-cheeked Bee-eater** and **Bee-eater** can be difficult. When confronted with a bee-eater species, the length of the central tail feathers (when present) and the colour of the throat and underparts are key.

Woodpeckers – p. 123
The presence or absence of red in the undertail is important when separating **Great Spotted** and **Lesser Spotted Woodpecker**, though a small woodpecker with lemon-yellow undertail-coverts can only be **Yellow-bellied Sapsucker**.

Falcons – p. 124
Ten species of falcon have been seen in Britain. The **Kestrels** can present a few difficulties, in particular when encountered in female plumage where face pattern is crucial. Pale claws will always positively identify both sexes of **Lesser Kestrel**, when separating this species from **Kestrel**. **Merlin** is the smallest of the falcons regularly seen. The males of **Red-footed** and **Amur Falcon** can be very similar in appearance and only the presence of clean white axillaries will clinch **Amur Falcon**. The females of all these falcons need good views of underpart colour and underpart/face pattern to separate them. Upperpart and underpart patterning, combined with the head pattern will help separate the larger falcons such as **Peregrine** and **Gyrfalcon.**

American flycatchers – p. 127
Six different North American flycatchers have been recorded in Britain. Two of these, **Eastern Phoebe** and **Eastern Kingbird**, are fairly straightforward to identify. Whilst three of these species (the *Empidonax* flycatchers: **Acadian**, **Yellow-bellied** and **Alder**) are altogether more difficult. Even North American birdwatchers familiar with the group often leave some unidentified. With good views and careful ob-

servation it is possible to reach an identification for most, though great care is needed in assessing colour saturation, strength of wingbars and eye-ring, and the length of the visible primaries beyond the tertials. **Great Crested Flycatcher** is under assessment but is included here.

Shrikes – p. 128
The shrikes come in two plumage colours: grey and brown. Brown plumage shrikes have brown or red tails whilst grey plumage shrikes have dark or black tails.
Female and immature brown plumage shrikes can be difficult to identify though a combination of tail shape, length and colour and the underpart patterning will help tease them apart. The immatures of **Woodchat** and **Masked Shrike**, whilst tricky, can be separated by the presence or absence of a white primary patch, with tail length and rump colour also of relevance. When separating the two grey shrikes (**Lesser Grey and Great Grey Shrike**) wing length and the size of the white primary patch are important.

Crows – p. 131
When seen well on the ground the black crows (**Jackdaw**, **Chough**, **Rook**, **Carrion Crow**, **Hooded Crow** and **Raven**) should present few problems. Tail shape is important when identifying birds in flight. **Hooded Crow** is illustrated here as a full species. It has been reclassified as a **Carrion Crow** subspecies, though this could well change again in the future.

Tits – p. 132
Marsh and Willow Tit present one of the most difficult identification challenges. The presence or absence of a pale spot in the bill and the extent of white in the cheek, combined with head shape and a dull or glossy crown will enable separation of the two, if good views are obtained.
Young **Blue** and **Great Tits** can be told apart by the colour of the crown alone. With good views, **Penduline, Bearded** and **Long-tailed Tits** should present few problems.

Larks – p. 134
Larks can be tricky and careful observation is needed. A combination of features will assist - these include the presence or absence of a

white trailing edge to the wing, tail length, outer tail feather colour, shape and size of any crest and underwing colour.

Swallows and Martins – p. 136
When seen well martins and swallows do not pose too many identification challenges. The presence or absence of a pale rump and collar, throat colour, undertail colour and tail shape are all key in aiding identification, along with the presence of any white tail spots.

Warblers

New World Warblers – p. 138
Nineteen species of **North American warbler** have been found in Britain. In general these form a rather bright and colourful group. This group can be split into two, those with strong markings, such as head and face pattern and wing-bars, against those that are relatively plain. Features to concentrate on in the first group are the colour and strength of the head pattern. Is there a crown-stripe? What colour are the supercilium and cheeks? How strong and what colour are the wingbars? Does it have a colourful rump and is there any pattern in the open tail? For those that are rather plain, the strength of the wingbars, if present, and the colouring or contrast between them can be very helpful. The strength of any upperpart and underpart streaking is important. Any colour in the throat and undertail should also be noted. With good views most should be straightforward to identify.

Old World Warblers – p. 141
For some of the Old World warbler species wing length (how far the primaries extend beyond the tertials on the folded wing) is an important feature for identification. The tertial/primary ratio illustrations are an average of measurements recorded in various bird ringing and identification guides and are to be used as a general guide as with some species there is overlap between the smallest of the large and the largest of the small.

Leaf warblers
There are fifteen *Phylloscopus* warblers on the British List, which includes the familiar **Willow Warbler** and **Chiffchaff** and rarities such as **Arctic** and **Radde's Warbler**. They are mostly varying shades of olive-green with some a bit more yellow, whilst others are browner.

The presence or absence of any wingbars, crown-stripes and underpart colouration are all helpful pointers. In common with all warblers, songs and calls are extremely helpful in clinching an identification. Wing length is also helpful - whether the length of the primary feathers extending beyond the tertials on the folded wing is short or long. For example, short in **Chiffchaff** and long in **Willow Warbler.**

Marsh warblers
The *Acrocephalus* warblers (8 species seen in Britain; these are referred to as 'reed and sedge warblers' but include other species e.g. **Aquatic, Blyth's Reed**) are mostly various shades of brown. They can be neatly split into two groups, un-streaked and streaked. Songs and calls are very helpful in achieving an identification, but critical features include wing length – whether the length of the primary feathers extending beyond the tertials on the folded wing is short or long – brightness and colour of the rump and the presence of a supercilium and crown stripe are also helpful.

Tree warblers
The four *Iduna* and *four Hippolais* warblers can be neatly split into the grey-brown *Idunas* (**Booted, Syke's and Eastern** and **Western Olivaceous Warblers**) and the olive-green *Hippolais* (**Melodious, Icterine** and **Olive-tree Warblers**). Songs and calls are very useful along with the presence of a pale wing panel, wing length (whether the length of the primary feathers extending beyond the tertials on the folded wing is short or long) and leg colour. A combination of these features should help identification.

Grass warblers
Of the five species of *Locustella* on the British List, three are streaked and spotted (**Pallas's Grasshopper, Lanceolated** and **Grasshopper Warbler**)and two are unstreaked (**River** and **Savi's Warbler**). However, except for **Pallas's Grasshopper Warbler**, all have marked undertail-coverts, spotted and streaked or pale-tipped. These separate them from the superficially similar *Acrocephalus* warblers.

Sylvia warblers
The *Sylvia* warblers (**Blackcap to Dartford Warbler**) are largely grey in colour and with sixteen different species on the British List can sometimes be problematic. However, many of them are well marked; useful

features include head pattern and colour, wing pattern, tail length, the amount and pattern of white in the outer tail and the colour of the underparts. The females are generally duller than the males and, in some species, can be quite difficult to separate. As with other warblers, wing length – the length of the primaries beyond the tertials on the folded wing – can be diagnostic. For example, in **Subalpine Warbler** the wing length is long, in **Sardinian** the wing length is short.

Unless the call is heard the females and immatures of **Moltoni's, Western** and **Eastern Subalpine Warbler** are extremely difficult to discern. To be sure the tail pattern also needs to be seen, in particular concentrate on the amount of white in the tip of T5 – the second to last tail feather working out from the middle of the tail. The only subspecies of **Lesser Whitethroat** illustrated is *blythi,* often referred to as Siberian Lesser Whitethroat, as this is currently the only one confirmed to regularly occur in Britain.

Thrushes – p. 155
There are seven species each of North American and eastern thrushes on the British List. All the American thrushes show a strong, pale bar on the underwing in flight, with the exception of **American Robin**, which has a plain underwing. As such it is not useful as a feature to separate them from each other and so not used as an identification feature here. Head pattern, the shape and strength of underpart spots and streaks and the colour of tail and rump can all be very helpful in securing an identification of the rarer thrushes and the common thrush species too.

Chats – p. 159
The **chats** are generally a rather colourful group and with good views most are readily identifiable, in particular the males. The females of this group can be challenging but breast pattern and tail colour is extremely useful in separating them.
Care is needed when separating the two **nightingales**, where a combination of wing structure, tail and rump colouring and the strength of any spots and streaks on the breast will help secure an identification.

Flycatchers – p. 161
The **flycatchers** can be split into two groups, those with rather plain wings and those with white patches in the wing. The key to separating the first group is the presence or absence of any breast streaking, the

tail pattern and rump colour, and in the case of **Brown Flycatcher**, bill colour. For the 'black-and-white' flycatchers (**Pied** and **Collared**), the extent of the small, white primary patch on the folded wing is key.

Redstarts – p. 162
Of the three **redstart** species ('starts') seen in Britain, the two seen most frequently (**Common** and **Black**) can be readily separated by the colour of the underparts. Only the male of the very rare **Moussier's Redstart** has occurred in Britain and can be identified as such by the strong head-pattern.

Chats – p. 163
Stonechats are a very challenging group to identify. A combination of the saturation of the colour of the underparts and the strength of the half-collar in the male is helpful. Rump colour and pattern (is the rump streaked or plain?) is also useful. As well as the tail pattern itself - the presence or absence of any white and how far it reaches down the tail – are also helpful in securing an identification but be aware that some may still be impossible to assign to a species. The females are extremely difficult to distinguish between, though rump and tail pattern are key if good views can be achieved. The eastern stonechats are so difficult to assign to a species that DNA analysis is often needed.
Whinchat is the only species in this group to show an extensive supercilium and heavily spotted rump.

Wheatears – p. 164
The seven species of wheatear on the British List pose a challenge, with female and immature plumages being particularly difficult at times. One of the key features to focus on is the proportion of white to black in the tail, which when seen well can help separate them. The colour of the throat and upperparts is also helpful. The western and eastern forms of **Black-eared Wheatear** can be impossible to separate in some female and immature plumages and even some out-of-range males may present some difficulty, some birds may have to be left as **Black-eared Wheatear** sp. An additional complication is found in the overlap zone (southeast France to Eastern Europe) between the two forms where intermediate birds are fairly common. As such, we have included **Black-eared Wheatear** as a single species with an eastern and western form.

Sparrows – p. 166
There are four sparrows (**House, Spanish, Tree** and **Rock**) and three accentors (**Dunnock, Alpine Accentor and Siberian Accentor**) on the British List. When considering **House** and **Spanish Sparrow** the colour and pattern of the crown, along with the extent of the 'bib' and underpart streaking, are important features in identifying the males, and separating these two species. The females of **House** and **Spanish Sparrow** are very difficult to separate from each other and at times this may prove impossible. **Tree** and **Rock Sparrow** are readily distinguished from each other, though the sexes for these species are alike. To help separate the three accentors on the British List, pay attention to the colour of the underparts.

Wagtails – p. 167
Although the nominate and several subspecies of **Yellow Wagtail** do sometimes occur in Britain, the only **Yellow Wagtail** that breeds in Britain is the subspecies (*flavissima*), which is illustrated.
For the other four species of wagtail, head pattern, upperpart and undertail colour are all useful pointers for identification and, apart from **Pied** and **White**, call is also helpful. To help separate **Pied** and **White Wagtail** the colour of the rump (black in **Pied** and grey in **White),** and the intensity of the grey wash on the flanks are key.

Pipits – p. 168
With eleven species on the British List, the pipits form a rather large group and what's more they are all very similar in appearance. However, with careful observation they can be identified. The colour of the lores (dark vs. pale), the intensity of upperpart and underpart streaking, the presence or absence of any pale 'tramlines' on the back, along with the head pattern and the length of the hind claw are all useful features to look for. **Pechora Pipit** is the only bird in the group that shows primaries beyond the tertials on the closed wing.

Finches – p. 170
The finches are a large, colourful group that should provide few challenges. The rump colour is often key to a correct identification. Other things to look for include the colour of upper and underparts, the presence or absence of wingbars, the colour of the outer tail feathers and the size and colour of the bill. When differing, the female plumage is usually a duller version of the male.

The **redpolls** have now been lumped into a single species – **Common Redpoll**. We have illustrated the previously full species (Common Redpoll. Lesser Redpoll and Arctic Redpoll) as distinct colour morphs. Brown morph for Lesser Redpoll, pale morph for Common Redpoll and white morph for Arctic Redpoll.

This group needs more care than most of the finches when teasing them apart. For the **redpolls** the ground colour, the background colour to the brown body feathering, is important. Is it generally pale/white, as in the **Common** and **Arctic Redpolls**, or does it have the more buff tone of the brown morph **Lesser Redpoll**? The colour and the extent of any streaking in the rump and undertail-coverts offer useful clues too.

Crossbills – p. 175

The presence and strength of white wingbars and tertial tips, along with the size and shape of the bill, are all important pointers when identifying the crossbills but be aware that some individuals can be difficult to assign to a species.

The heads of **Scottish Crossbill** have been illustrated but there is reasonable doubt that it is a full species. Its taxonomic status is currently not fully understood and it may well prove to be a subspecies of **Common** or **Parrot Crossbill.** Future research may well conclude Common, Parrot and Scottish are actually a single species.

Buntings – p. 176

There are 19 species of bunting on the British List and as a group they can be challenging to identify. The winter and female plumages are often more difficult to assign to a species than the summer males. The head pattern, rump colour and the strength and colour of any underpart streaking are all important features to look for but be cautioned that without good views some may remain unidentified.

New World Sparrows – p. 180

North American sparrows often resemble Old World buntings. Pay attention to the head pattern, and the colour and the extent of any colour in the rump and tail. The colour and strength of any underpart streaking is also an important aid to identification, as is the presence or absence of pale or white outer tail feathers.

Very Recent Additions — p. 182

The final plates in the book shows all the species that have been added to the British List since the book was started. There have been a few that were added before the plate they appear on was put together and we were able to fit these in as we were working on them.

The first **Cape Gull** was originally identified as a **Kelp Gull**, possibly of the subspecies *vetula*, which is commonly known as **Cape Gull**. As the bird aged and retained a dark eye it became clear that this was indeed the case.

Madeiran Petrel is interesting in that the bird wasn't actually seen by anybody, a tagged bird was tracked into British waters and currently constitutes the only record of this species.

Thanks to advancements in digital camera systems, Britain's one and only **Soft-plumaged Petrel** was confirmed as such by a set of excellent photographs and video.

Red-rumped Swallow has recently been split into two species with **Western** and **Eastern Red-rumped Swallow** both on the British List.

Booted Eagle has had a chequered past in Britain with several claims rejected. However, during the writing of this book two wide-ranging individuals have been well documented. Rather conveniently one was a pale morph and the other a dark morph. A good pointer for both is the white 'headlights' that can be seen either side of the head and the pale 'window' in the wing when seen from below.

Red-legged Partridge 🟢🟡🟠🔵

- red
- white
- black necklace
- black edged chestnut bars
- red

Grey Partridge 🟢🟡🟠🔵

- strongly patterned upperparts
- orange
- chestnut bars
- dark chestnut patch
- green

Quail 🟢🟡🟠

- long pale supercilium
- black throat (♂)
- pale
- strongly streaked
- short tail
- chestnut

Corncrake 🟢🟡🟠

- pink
- brown stripe behind eye
- blue-grey
- chestnut
- chestnut, white, and black barring

Pheasant 🟢🟡🟠🔵

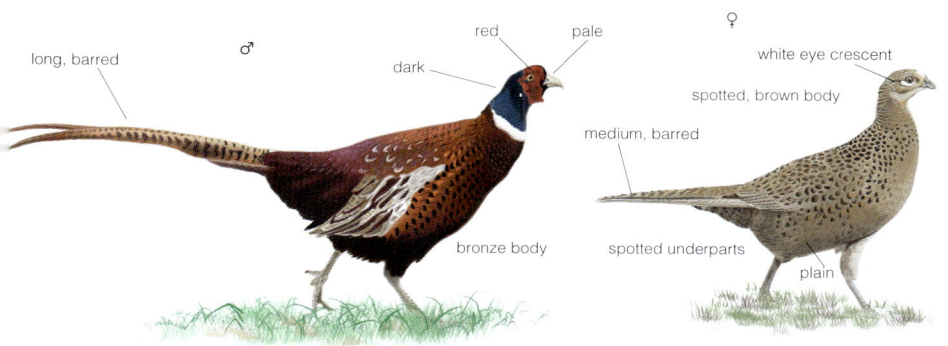

Golden Pheasant 🟢🟡🟠🔵

Lady Amherst's Pheasant 🟢🟡🟠🔵

Brent Goose 🟢🟡🟠🔵

Red-breasted Goose 🟢🟠🔵

Canada Goose 🟢🟡🟠🔵

- dark
- white chinstrap
- black
- pale-brown body
- pale

Cackling Goose 🟢🟠🔵

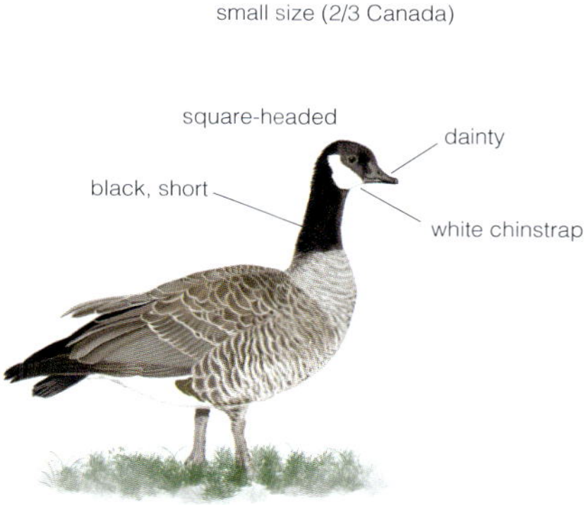

small size (2/3 Canada)

- square-headed
- black, short
- dainty
- white chinstrap

Barnacle Goose 🟢🟡🟠🔵

- grey-and-black barring on forewing

- dark, small
- white
- black
- grey-and-black barring
- white

Snow Goose 🟢🟠🟣

dark morph 
- white
- orange, small
- variable dark
- distinctive dark 'grin patch'
- dark upperparts

white morph
- orange, small
- distinctive dark 'grin patch'
- all-white body
- black flight feathers
- bright pink

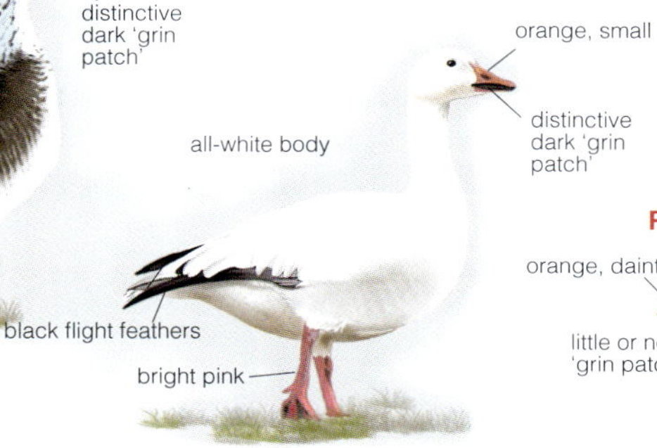

Ross's Goose 🔵

- orange, dainty
- little or no 'grin patch'
- all-white body
- bright pink
- black flight feathers

33

Greylag Goose 🟢🟡🟠🔵

Taiga Bean Goose 🟠🟢🔵

Tundra Bean Goose 🟢🟠🔵

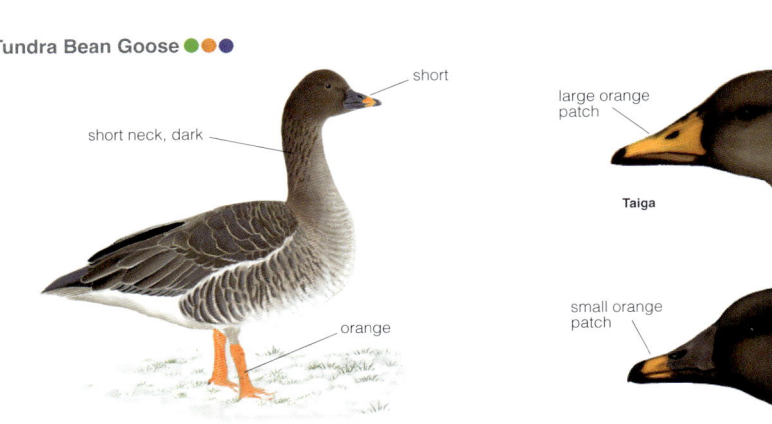

Pink-footed Goose 🟢🟡🔵

White-fronted Goose 🟢🟠🔵

White-fronted Goose subspecies

flavirostris (Greenland)

albifrons (NE Europe)

Lesser White-fronted Goose 🟢🟠🔵

36

Mute Swan 🟢🟡🟠🔵

- orange wash
- black mask
- black 'knob'
- orange

Ad.

Imm.

long tail

Bewick's Swan 🟢🟠🔵

- small yellow patch
- mostly dark
- call: high-pitched yodelling
- short neck

Ad.

Imm.

short tail

Whooper Swan 🟢🟡🟠🔵

- long neck
- dark tip
- large yellow patch - reaches beyond nostril

Ad.

call: deep bugling

Imm.

short tail

Bewick's Swan Ad.

Whooper Swan Ad.

Egyptian Goose 🟢🟡🟠🔵

- dark-rufous eye-patch
- pale sandy-brown body
- white forewing – visible in flight and at rest
- rich rufous
- dark pink

Ruddy Shelduck 🟡

- pale brown
- rich cinnamon body
- black
- white forewing – visible in flight and at rest
- black

Shelduck 🟢🟡🟠🔵

- white body
- dark green
- red
- broad chestnut band
- pink

Mandarin Duck 🟢🟡🟠🔵

♂
- orange sails
- distinct head pattern
- pink/red
- purple/dark
- two vertical white stripes

♀
- white eye-ring and adjoining white stripe behind eye
- white nail on end of bill
- dull pink
- white
- large pale spots

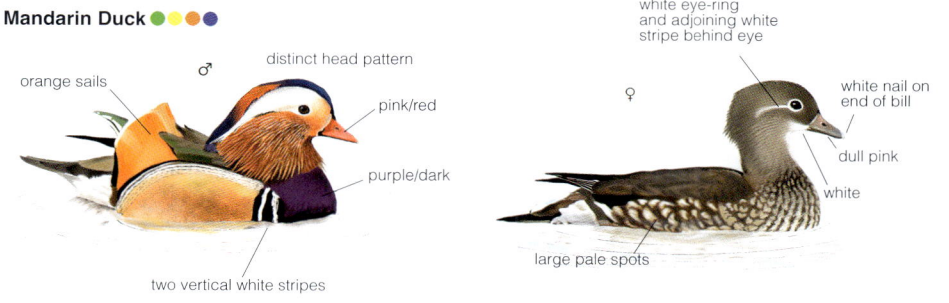

38
Baikal Teal 🟢🟠🔵　　　　　　　　　　Males

- long scapulars
- yellow face with black teardrop
- vertical white stripe
- black
- white trailing edge

Garganey 🟢🟡🟠
- long scapulars
- white crescent
- pale grey/blue forewing
- two broad white wingbars
- pale grey

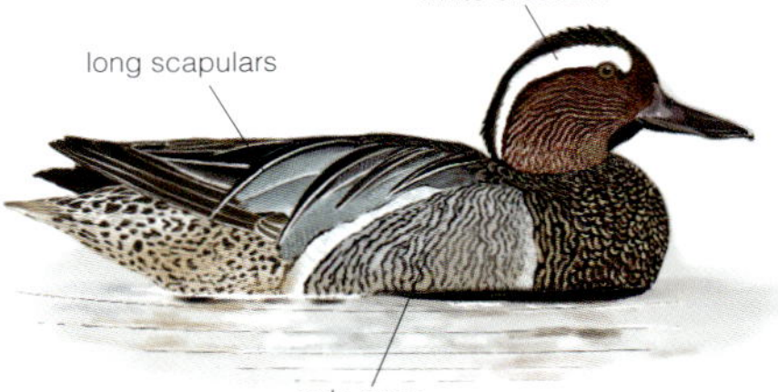

Blue-winged Teal 🟢🟡🟠🔵
- white vertical band
- spotted breast and flanks
- blue forewing
- green speculum
- white oval

Teal 🟢🟡🟠🔵
- white horizontal stripe
- distinct head pattern – strong buff borders
- bright green speculum
- yellow
- bright green
- mid-grey

Green-winged Teal 🟢🟡🟠🔵
- yellow
- distinct head pattern – weak buff borders
- mid-grey flanks
- vertical white stripe
- bright green speculum

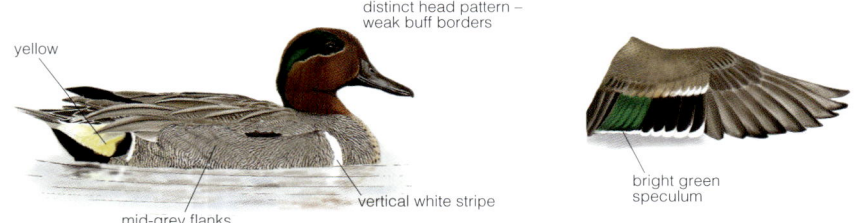

Gadwall 🟢🟠🟠🟣

- grey/black
- ♂
- white speculum
- black rear end
- pale/white
- yellow/orange
- white speculum
- ♀
- dark tertials
- orange cutting edge along bill
- finely speckled
- white speculum
- yellow-orange

Mallard 🟢🟡🟠🟣

- metallic green head
- yellow
- ♂
- two black curls on uppertail
- blue, white-edged speculum
- orange
- pale supercilium
- ♀
- pale tertials
- dark, smudged yellow
- blue, white edged speculum
- orange

Black Duck 🟢🟡🟠🔵

Pintail 🟢🟡🟠🔵

Wigeon 🟢🟡🟠🟣

Males

- buff/orange
- chestnut
- blue-grey, edged black
- large white panel
- grey

American Wigeon 🟢🟠🟣

- large white panel
- green blaze
- cream/white
- blue-grey, edged black
- warm brown

Falcated Duck 🟢🟣

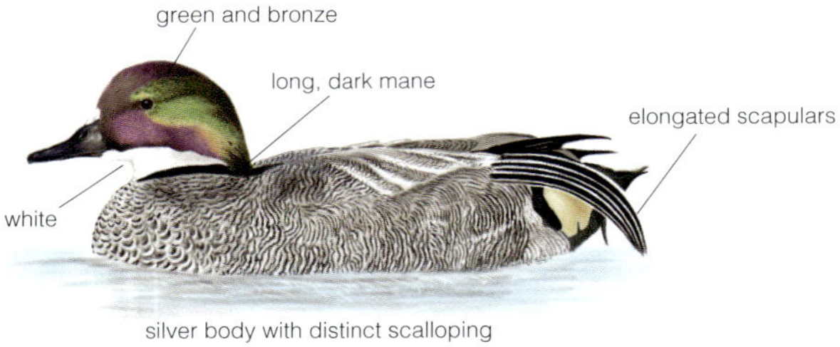

- green and bronze
- long, dark mane
- elongated scapulars
- white
- silver body with distinct scalloping

Shoveler 🟢🟡🟠🟣

- blue forewing
- dark green
- long spatulate
- chestnut
- white body

Wigeon 🟢🟡🟠🔵

Females

brown, speckled black

plain wing

blue-grey, edged black

breast paler than head

dull orange

American Wigeon 🟠🟡🔵

broad whitish wingbar

grey-brown, speckled black

small, dark eyepatch

blue-grey, edged black

rich orange

Falcated Duck 🟠🔵

finely streaked dark

slightly maned

all dark

buff line to edge of uppertail

dark chevrons

Shoveler 🟢🟡🟠🔵

short neck

grey-blue forewing

long spatulate

green speculum

orange

Pochard 🟢🟡🟠🟣

Canvasback 🟢🟡🟣

Redhead 🟢🟡🟣

Red-crested Pochard 🟢🟡🟠🟣

Ruddy Duck 🟡🟠🟣

Scaup 🟢🟡🟠🔵

Females

- white wingbar
- brown
- yellow/orange eye
- white face
- grey/brown back and flanks

Lesser Scaup 🟢🟠🔵

- peaked hind crown
- yellow eye
- white face
- tiny black nail
- grey/brown back and flanks
- grey outer and white inner wingbar

Tufted Duck 🟢🟡🟠🔵

- short tuft
- dark brown
- yellow eye
- white wingbar (more obvious on inner wing)
- mid-brown

- peaked hind crown
- pale spot
- white eye-ring and eye-stripe
- subterminal white band

Ring-necked Duck 🟢🟡🟠🔵

- grey wingbar

Ferruginous Duck 🟢🟡🟠🔵

- dark eye
- faint dark neck ring
- grey, tipped black, long
- broad white wingbar (almost to tip of wing)
- white

Eider 🟢🟡🟠🟣

Females

- narrow white wingbar
- pale tipped, wedge-shaped
- green/grey inner bill
- mid-brown, barred black
- heavily barred

King Eider 🟢🟡🟠🟣

- two small sails on back
- pale spot
- pale band behind tip
- rusty-brown with dark chevrons
- two narrow wingbars

Steller's Eider 🟢🟡🟠🟣

- square head
- dark-brown body
- pale eye-ring
- grey
- two broad white wingbars

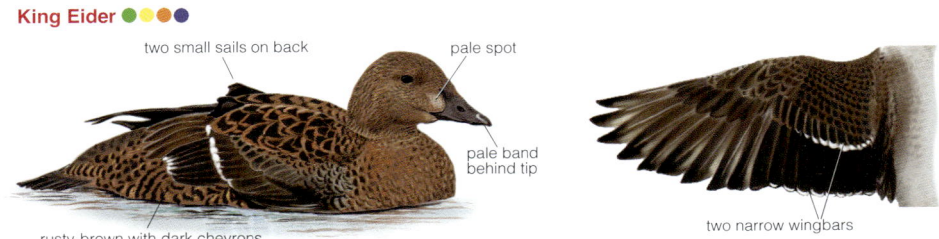

- dark cap
- round head
- pale face
- long

Common Scoter 🟢🟡🟠🟣

- pale primaries

Velvet Scoter 🟢🟡🟠🟣

- pale patch
- pale spot
- slightly concave
- medium length
- white secondaries

Surf Scoter 🟢🟡🟠🟣

Males

White-winged Scoter 🟢🟡🟠🟣

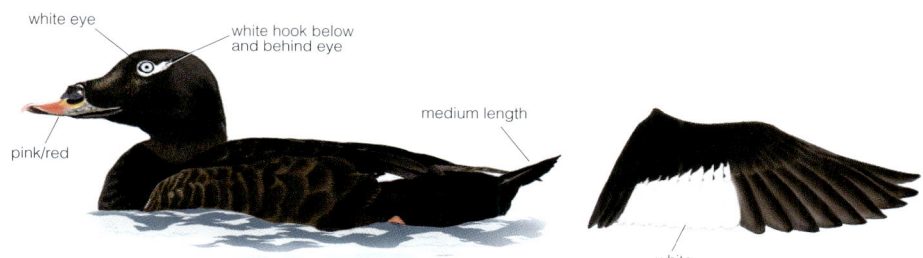

Black Scoter 🟢🟡🟠🟣

Surf Scoter 🟢🟡🟠🔵

Females

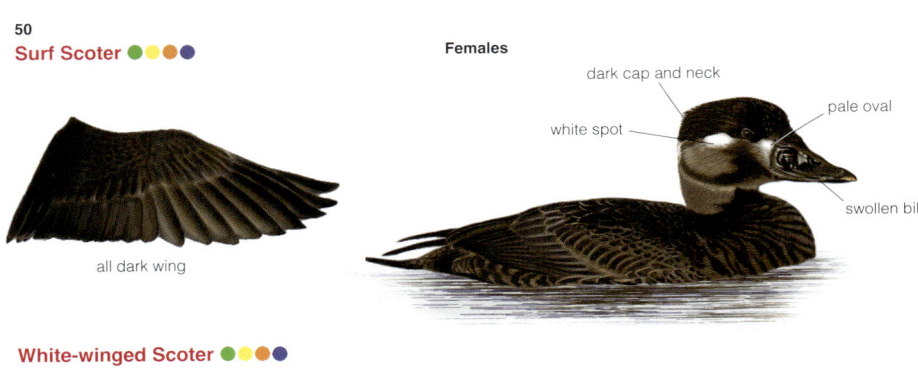

White-winged Scoter 🟢🟡🟠🔵

Black Scoter 🟢🟡🟠🔵

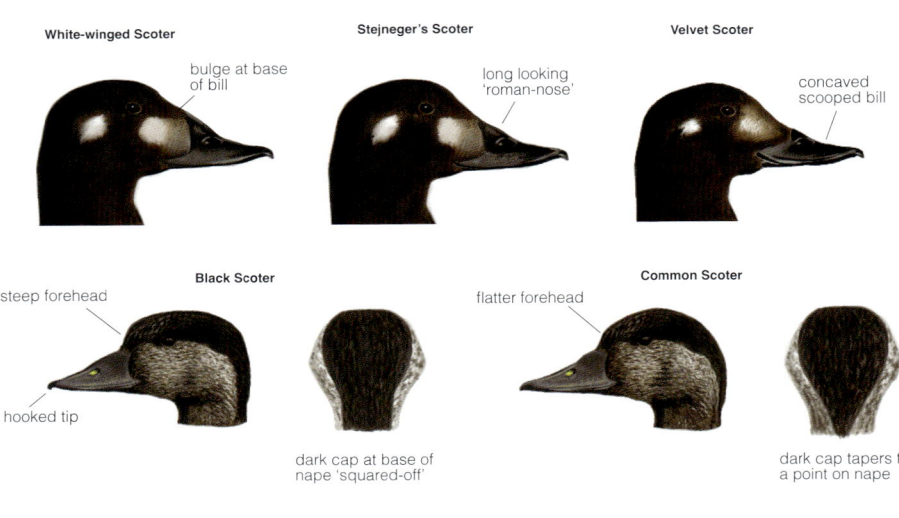

Harlequin 🟢🟡🟠🔵

Long-tailed Duck 🟢🟡🟠🔵

Goosander 🟢🟡🟠🔵

Males

- green gloss
- black
- red, long, thin
- white/peach
- large white patch on inner wing

Red-breasted Merganser 🟢🟡🟠🔵

- red eye
- red, long, thin
- green gloss
- orange-brown, streaked black
- two black bars in white inner wing

Hooded Merganser 🟢🟡🟠🔵

- prominent black-and-white crest
- mustard-yellow eye
- grey, long, thin
- orange
- patchy grey-and-white wing

Smew 🟢🔵

- white oval on forewing
- white head and neck
- black
- black stripe
- finely vermiculated grey

Goosander 🟢🟡🟠🟣

Females

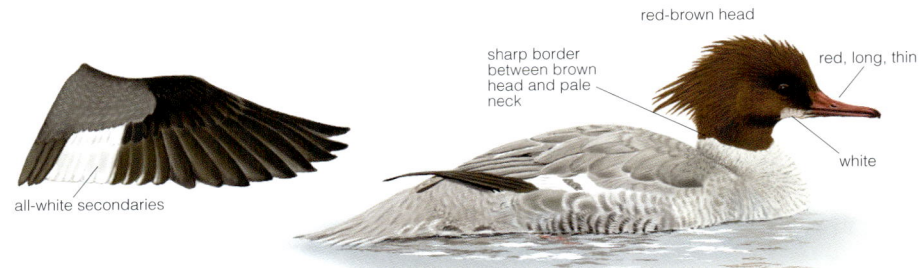

Red-breasted Merganser 🟢🟡🟠🟣

Hooded Merganser 🟢🟡🟠🔴

Smew 🟢🟣

Common Nighthawk ●

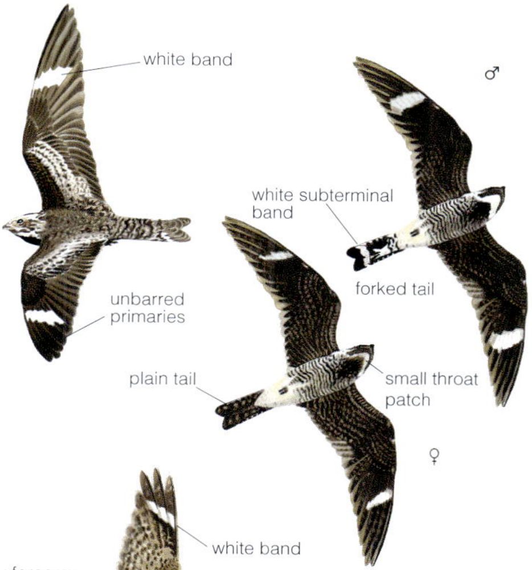

white/pale throat · white band · unbarred primaries · plain tail · white subterminal band · forked tail · small throat patch · ♂ · ♀

Red-necked Nightjar ●

long · rufous · grey forearm · white · white band

Nightjar ●●

black/dark forearm · large white spots · mid-length · brown barring · barred primaries · white band · white · ♂ · ♀

Egyptian Nightjar ●

pale sandy body · short · black bars · black · pale underwing

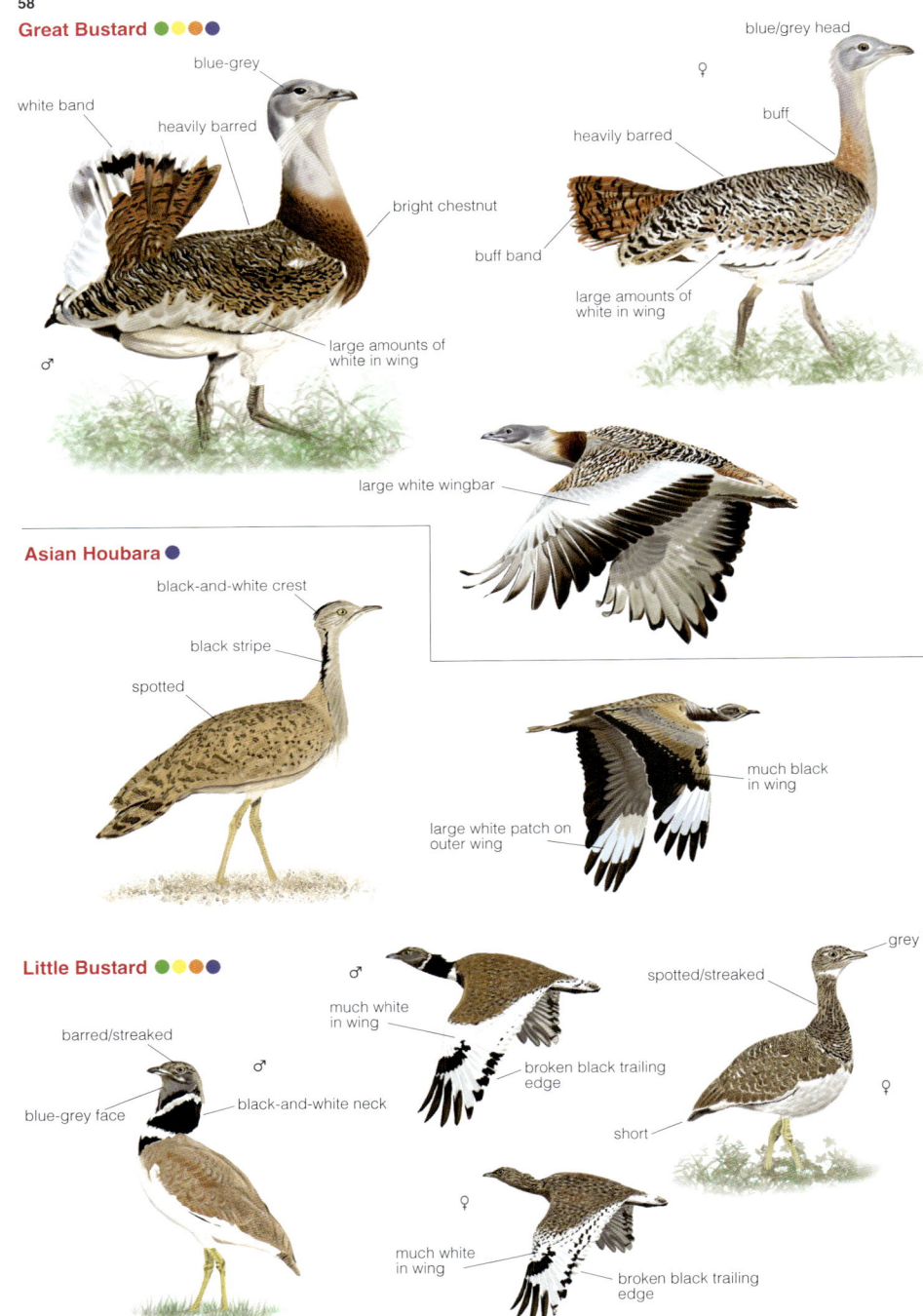

Turtle Dove 🟢🟡🟠

Oriental Turtle Dove 🟢🔵🔴

Pallas's Sandgrouse 🔴🟡🔵🟢

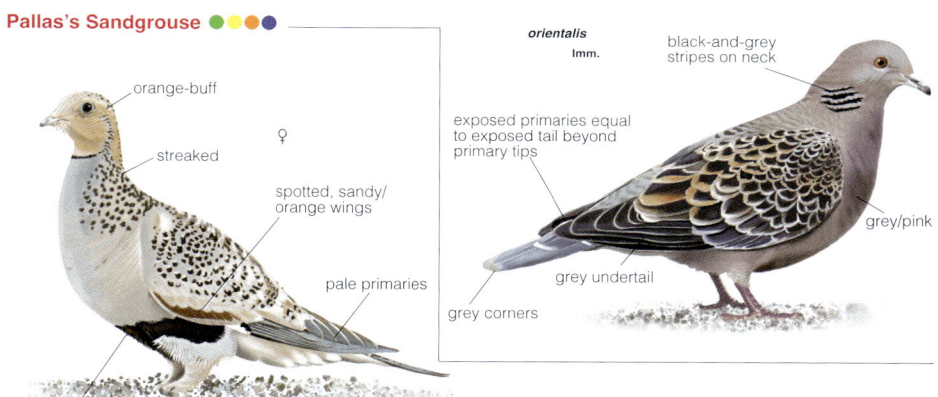

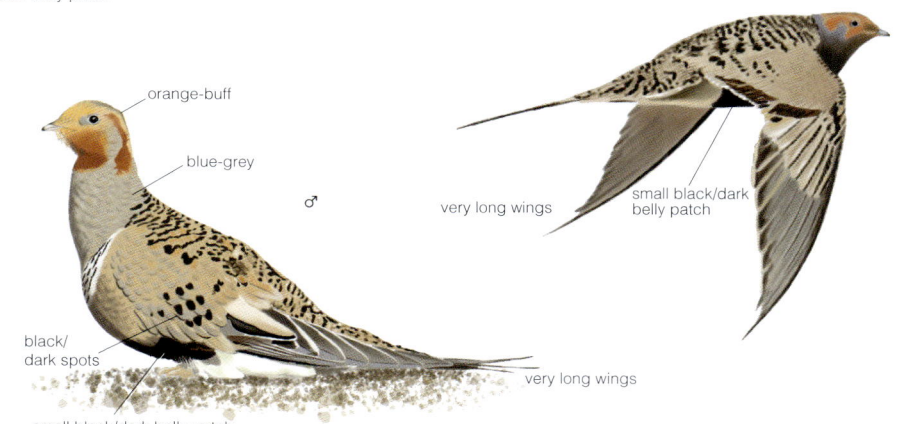

64

Moorhen 🟢🟡🟠🔵

Ad.
- red
- yellow tip
- blue-black body
- black centre to white undertail
- broken white line

Imm.
- mid-brown-grey body
- pale
- dull orange-red
- black centre to white undertail
- broken white line

Coot 🟢🟡🟠🔵

Ad.
- round edge to top of shield
- white shield and bill
- black 'tick' into lower part of shield
- black body
- dark undertail

Juv.
- grey body
- dark undertail
- white spot on lores
- grey
- pale

American Coot 🟢🟠🔵

1st-winter.
- small, dull red callus on frontal shield
- ivory-grey
- straight edge to shield/bill feathering
- dark bar across tip

- white undertail split by black bar

Stone-curlew 🟢🟡🟠🔵

- yellow eye
- yellow, black tip
- black-and-white bars
- yellow
- distinctive black-and-white wing pattern
- broad white wingbar

Oystercatcher 🟢🟡🟠🔵

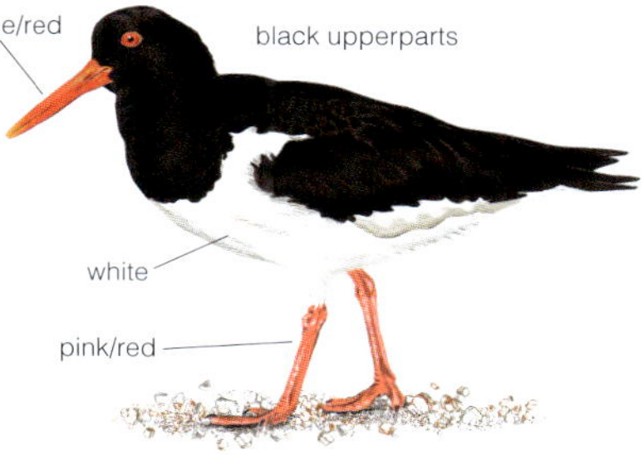

- orange/red
- black upperparts
- white
- pink/red

Black-winged Stilt 🟢🟡🟠🔵

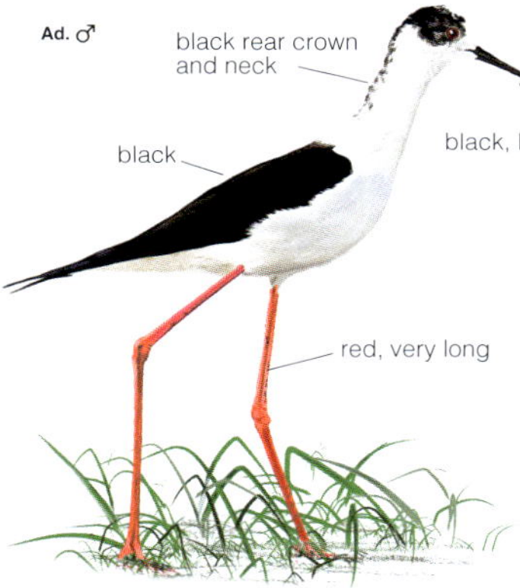

Ad. ♂
- black rear crown and neck
- black
- black, long, thin, straight
- red, very long

Ad. ♀
- white/dusky rear neck
- brown/black
- black, long, thin, straight
- red, very long
- black wings

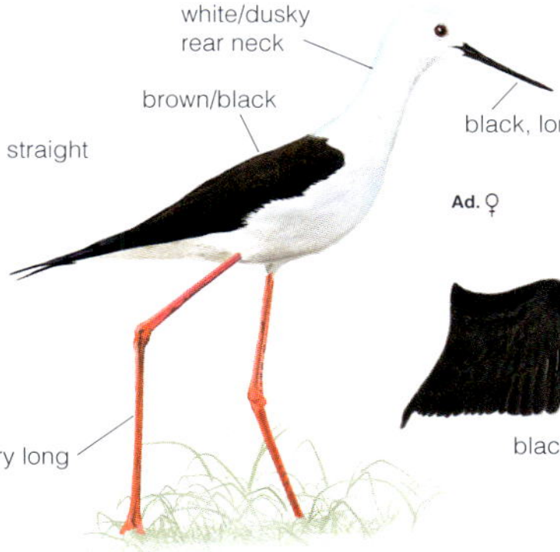

Avocet 🟢🟠🔵

- two black patches on wing
- black
- green/grey
- dark, thin, upturned
- black-and-white wings

Lapwing 🟢🟡🟠🟣

- black crest
- dark green
- black band
- chestnut
- very broad wings, tipped grey
- black tail band

Sociable Lapwing 🟢🟠🟣

- distinctive black-and-white head pattern
- sandy-grey body
- streaked
- yellow wash meets on back of head
- pale fringes to back and wing feathers
- streaked
- white belly
- Ad.
- dark lower belly patch
- black tail band
- 1st-winter
- distinctive wing pattern – sandy-grey, white, and black

White-tailed Lapwing 🟢🟡

- plain sandy-brown head and body
- grey/brown
- very long, bright yellow
- all white
- distinctive wing pattern, white reaches front and back of wing

Grey-headed Lapwing 🟢

- grey-brown, white, and black wings
- yellow, black tip
- grey
- black
- long, yellow

Ringed Plover 🟢🟡🟠🟣

- round lower edge to black cheek
- primaries visible beyond tertials
- **summer**
- orange, tipped black, stout
- orange
- **winter**
- primaries visible beyond tertials
- cream supercilium
- dark, stout
- round lower edge to brown cheek
- orange
- white wingbar
- single web between toe

Semipalmated Plover 🟢🟡🟠

- short, or no pale supercilium
- **Ad. summer**
- orange, tipped black, short, stubby
- call: disyllabic 'cheewi'
- thin black band
- webs between middle and inner toes
- dark line separating supercilium from cheek patch
- dark, short, stubby
- narrow white streak above the gape, reaches the upper mandible
- **1st-winter**

Little Ringed Plover 🟢🟡🟠

- white band behind dark forehead band
- long tertials - no primary projection
- yellow orbital ring
- **Ad. summer**
- yellow/pink
- very faint wingbar
- dull yellow orbital ring
- dark, slender
- dull buff supercilium
- long tertials – no primary projection
- pointed lower edge to dark cheek
- **1st-winter**

Killdeer 🟢🟡🟠

- **summer**
- dark, long
- double black breast-band
- long
- rufous rump and uppertail
- prominent, broad white wingbar

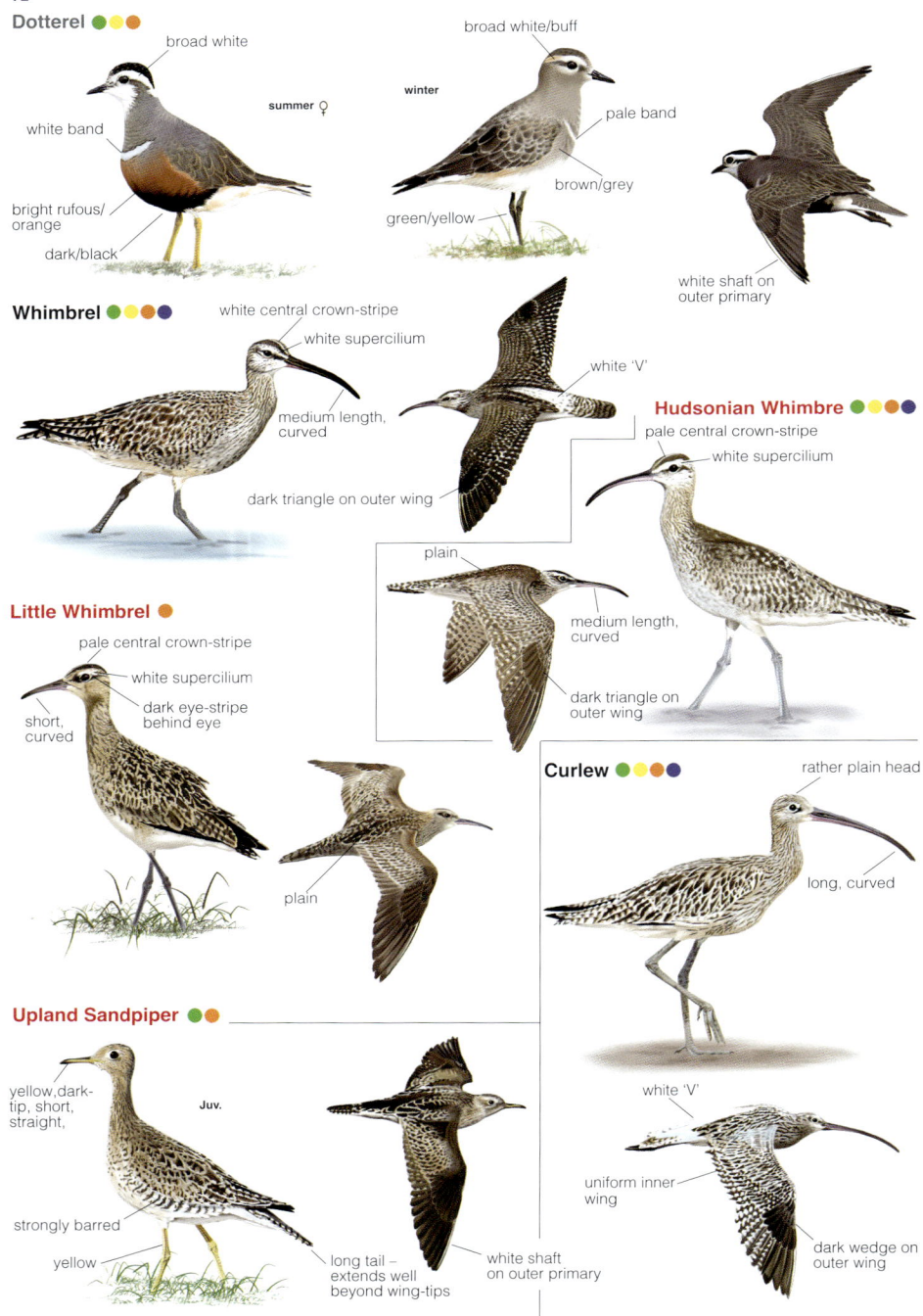

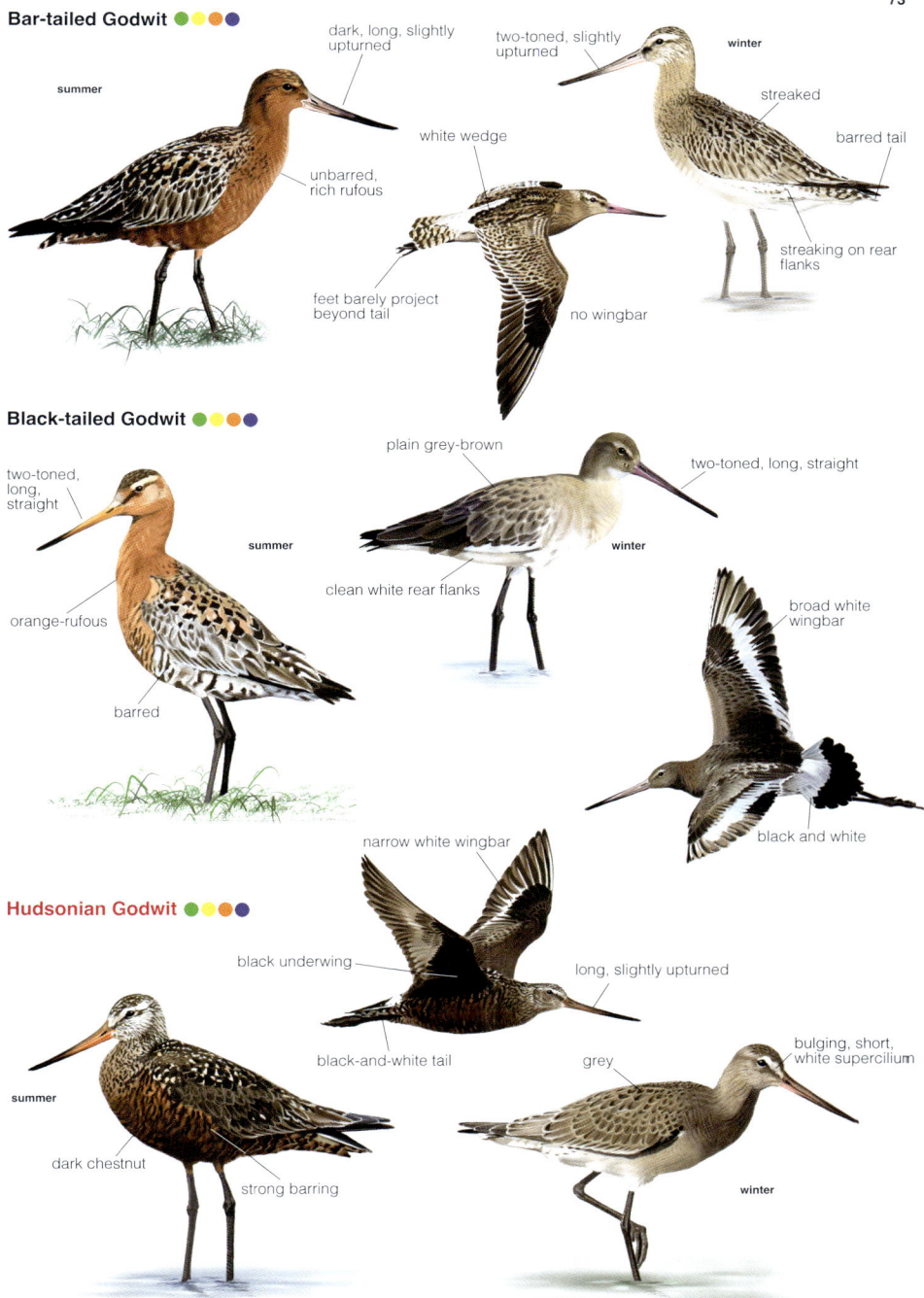

Sharp-tailed Sandpiper 🟠🔴

Ad. summer
- bright rufous, streaked crown
- dark 'V's
- obvious markings on flanks
- green
- breast markings fade out on lower breast

Juv.
- pale supercilium
- thin white bars
- rufous, streaked
- streaked necklace
- plain, with peachy-orange wash

Pectoral Sandpiper 🟡🟠

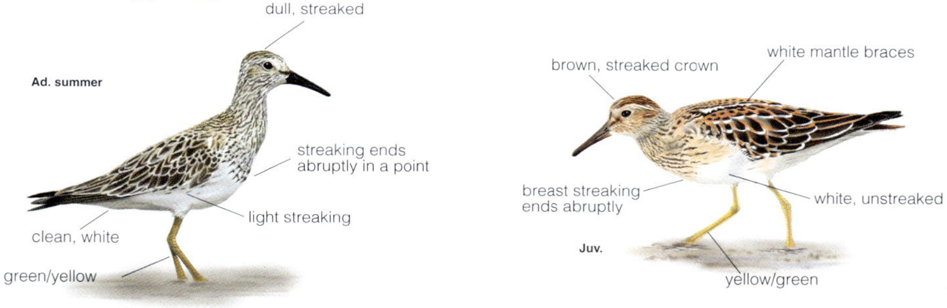

Ad. summer
- dull, streaked
- streaking ends abruptly in a point
- light streaking
- clean, white
- green/yellow

Juv.
- brown, streaked crown
- white mantle braces
- breast streaking ends abruptly
- white, unstreaked
- yellow/green

Ruff 🟢🟠🟡🔵

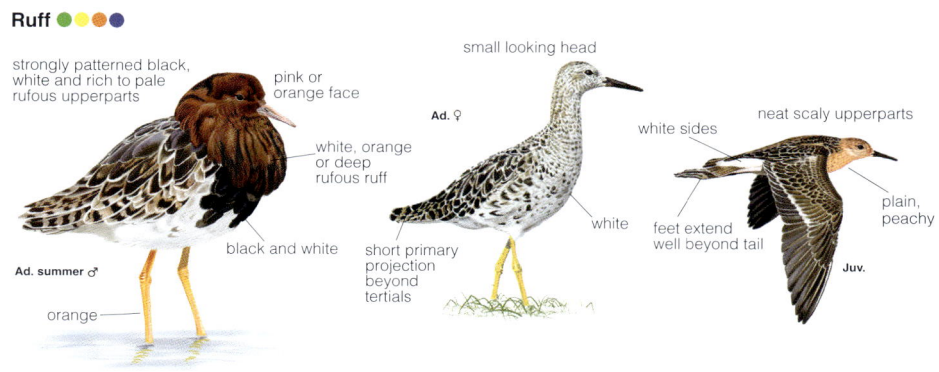

Ad. summer ♂
- strongly patterned black, white and rich to pale rufous upperparts
- pink or orange face
- white, orange or deep rufous ruff
- black and white
- orange

Ad. ♀
- small looking head
- white
- short primary projection beyond tertials

Juv.
- neat scaly upperparts
- white sides
- feet extend well beyond tail
- plain, peachy

Buff-breasted Sandpiper 🟡🟠

Ad. summer
- long primary projection beyond tertials
- buff
- yellow/mustard

Juv.
- large eye
- spots on side of upper breast
- neat, scaly upperparts
- dark crescent on white underwing
- plain uppertail

Woodcock 🟢🟡🟠🟣

- transverse bars
- pale/grey bars
- bright rufous
- silver-grey band
- deep belly in flight

Jack Snipe 🟠🟢

- dark crown
- golden yellow bars
- cheek bar joins eye-stripe
- streaked
- rather plain tail

Great Snipe 🟢🟠

- white bars on wing
- white wingbar
- heavily barred underparts
- white sides to tail
- dark plain underwing
- white sides to tail

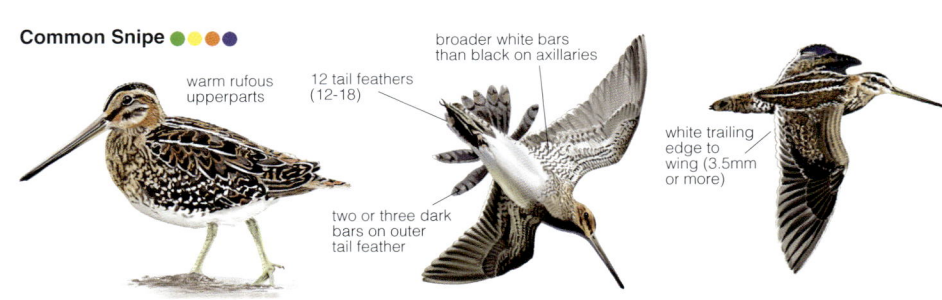

Common Snipe 🟢🟡🟠🟣

- warm rufous upperparts
- 12 tail feathers (12-18)
- broader white bars than black on axillaries
- two or three dark bars on outer tail feather
- white trailing edge to wing (3.5mm or more)

Wilson's Snipe 🟠

- cold, monochrome upperparts
- 16 tail feathers (14-18)
- broader black bars than white on axillaries
- four or more dark bars on outer tail feather
- narrow white trailing edge to secondaries (2mm or less)

Wood Sandpiper 🟢🟠🔵

- prominent supercilium
- white/silver spangled upperparts
- narrow bars
- long yellow-green tibia
- Ad.

- prominent supercilium
- yellow/golden spotted upperparts
- diffuse edge
- long yellow-green tibia
- Imm.

- white

Marsh Sandpiper 🟢🟠🔴🔵

- hardly any primary projection beyond tail
- black spots on upperparts
- Ad. summer
- very fine

- obvious cap
- very fine
- plain grey-brown upperparts
- Ad. winter

- white wedge
- long legs project beyond tail

- hardly any primary projection beyond tail
- pale fringes to upperparts
- very fine
- Imm.

Greenshank 🟢🟡🟠🔵

- dark spots on upperparts
- sturdy, upturned
- green
- Ad. summer

- white wedge
- dark outer wing

- pale fringes to upperparts
- sturdy, upturned
- green
- Imm.

- pale head
- grey upperparts
- sturdy, upturned
- Ad. winter

Terek Sandpiper 🟢🟠🟡🔵

- steep forehead
- upturned
- black bar on upperparts
- Ad.

- steep forehead
- upturned
- pale fringes to upperparts
- Imm.

- white trailing edge
- grey tail and rump

Kittiwake 🟢🟡🟠🟣

- black tip
- Ad. winter
- contrasting silvery outer wing
- black shawl
- black
- Imm.
- yellow, drooping
- dark, short
- little or no white spotting on folded wing
- Ad. summer
- black, short
- tiny white spots on folded wing
- Imm.
- black bar across middle wing, forms black 'W' across both wings
- grey-brown, white, and black wings

Sabine's Gull 🟡🟠

- dark grey
- grey, white, and black wings
- white tips on long wings
- dark, yellow tip
- white tips on long wings
- dark
- Ad. summer
- dark shawl
- Ad. summer
- Imm.
- largely dark head
- Ad. winter
- grey, white, and black wings
- scaly upperparts
- all dark, slender
- Imm.
- grey
- narrow white trailing edge
- black bar on wing, forms black 'W' across both wings
- Ad. summer
- black/dark underwing
- white trailing edge
- dark cap, black spot behind eye

Little Gull 🔵🟢🟡🟠

- narrow white trailing edge
- black bar
- grey nape
- Ad. summer
- black
- dark cap, black spot behind eye
- Imm.
- dark, tiny
- Ad. winter
- red
- broad dark bars on wing, forms black 'W' across both wings
- pale cap
- white trailing edge

Ross's Gull 🟣🟢🟡🟠

- black neck ring
- dark, small
- grey wings
- Ad. winter
- dark, small
- Ad. summer
- pink flush
- broad white trailing edge
- wedge-shaped
- small dark patch around eye
- Imm.

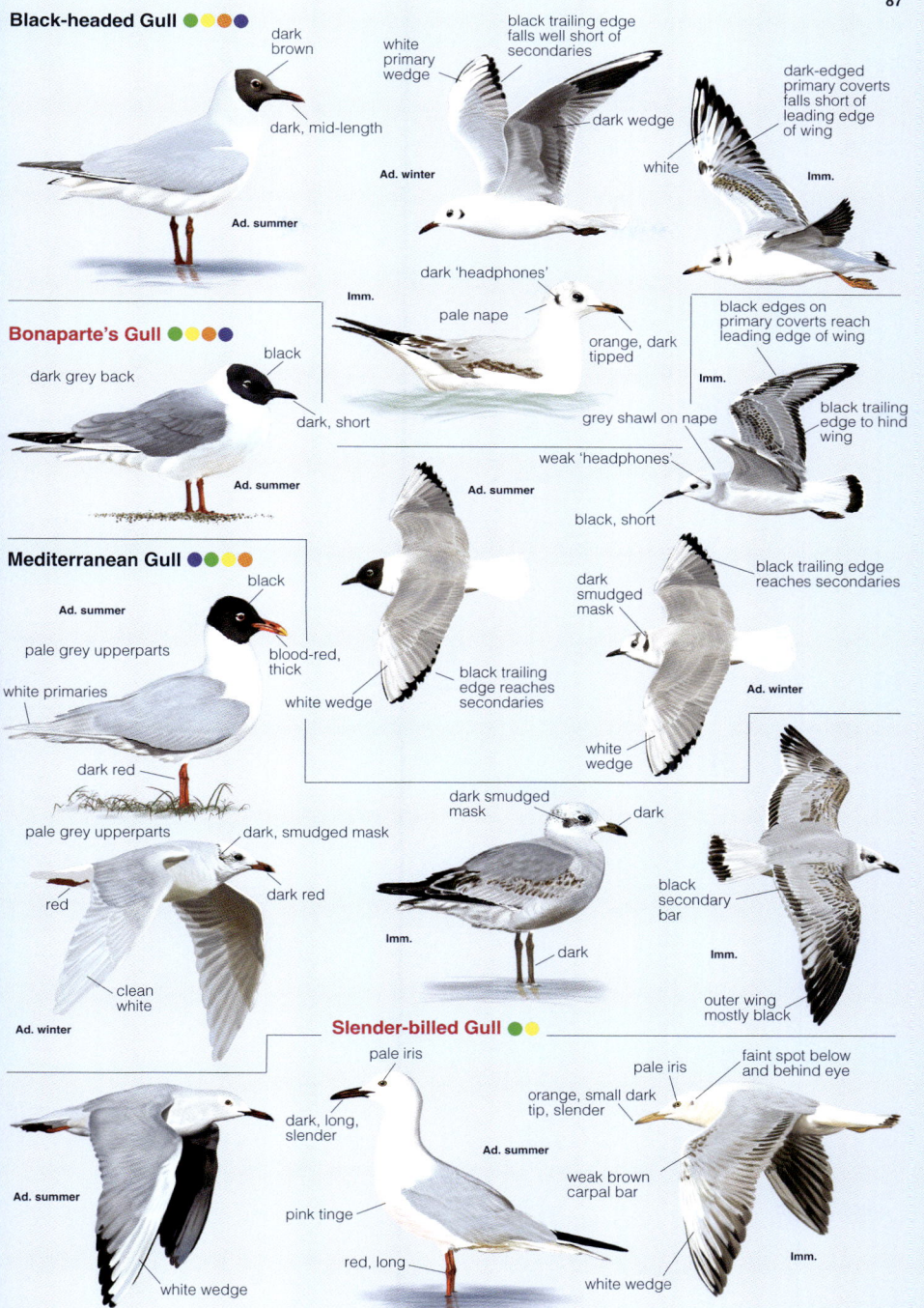

Laughing Gull 🔴🟡🟢🔵

Franklin's Gull 🔴🟡🟢🟣

Common Gull 🔴🟢🟡🟠

Ring-billed Gull 🔵🟢🟡🔴

Herring Gull 🟠🟡🟢🟣

Ad. summer
- yellow, stout
- yellow eye
- pale grey upperparts
- pink

Ad. winter
- strongly streaked
- yellow eye
- yellow, stout
- pink
- large, white 'mirror' spot

Juv.
- large pale 'window' (inner primaries)

2nd-winter
- pale eye
- dark sub-terminal bands on some mantle feathers
- short wings - don't extend much beyond tail
- pale pink base, extensive dark tip
- evenly spotted greater coverts

Juv.
- all dark, stout
- pink
- evenly spotted greater coverts
- 'holly leaf' patterned tertials

American Herring Gull 🟠🟢🟣

Juv.
- brown head and body
- all-dark tail
- heavily barred undertail
- dark rump
- small pale 'window' in inner primaries
- pink, dark tip
- extensive dark smudging on underparts

2nd-winter
- almost all dark
- solid brown panel forms dark bar in middle of wing
- dark tips to inner primaries

Yellow-legged Gull 🟢🟡🟠🟣

Ad. summer
- mid-dark grey upperparts
- yellow, strong
- yellow

Ad. winter
- fine brown limited streaking
- yellow, strong
- solid black wing-tips with short grey tongues
- small, white 'mirror' spot

Juv.
- pale face
- dark eye mask
- pink, long
- dark-centered tertials with pale edges

2nd-winter
- largely white head
- narrow dark subterminal bars to some mantle feathers
- long-winged
- broad white tips to tertials

Caspian Gull 🟠🟡🟢🟣

Ad. summer
- small looking pear-shaped head
- small dark eye
- yellow, long

Ad. winter
- grey tongues extend well into black wing-tip (p6-p8)
- large, white 'mirror' spot

Juv.
- dark double wingbarred look (formed by dark secondaries and dark bases to greater coverts)
- broad rectangle on greater coverts
- dark pale fringed tertials with small dark tips
- long-winged
- fairly plain greater coverts

2nd-winter
- pale pear-shaped head
- dark 'anchor' shapes in grey mantle feathers
- often has small white spot on wing-tip
- long-winged
- fairly plain greater coverts

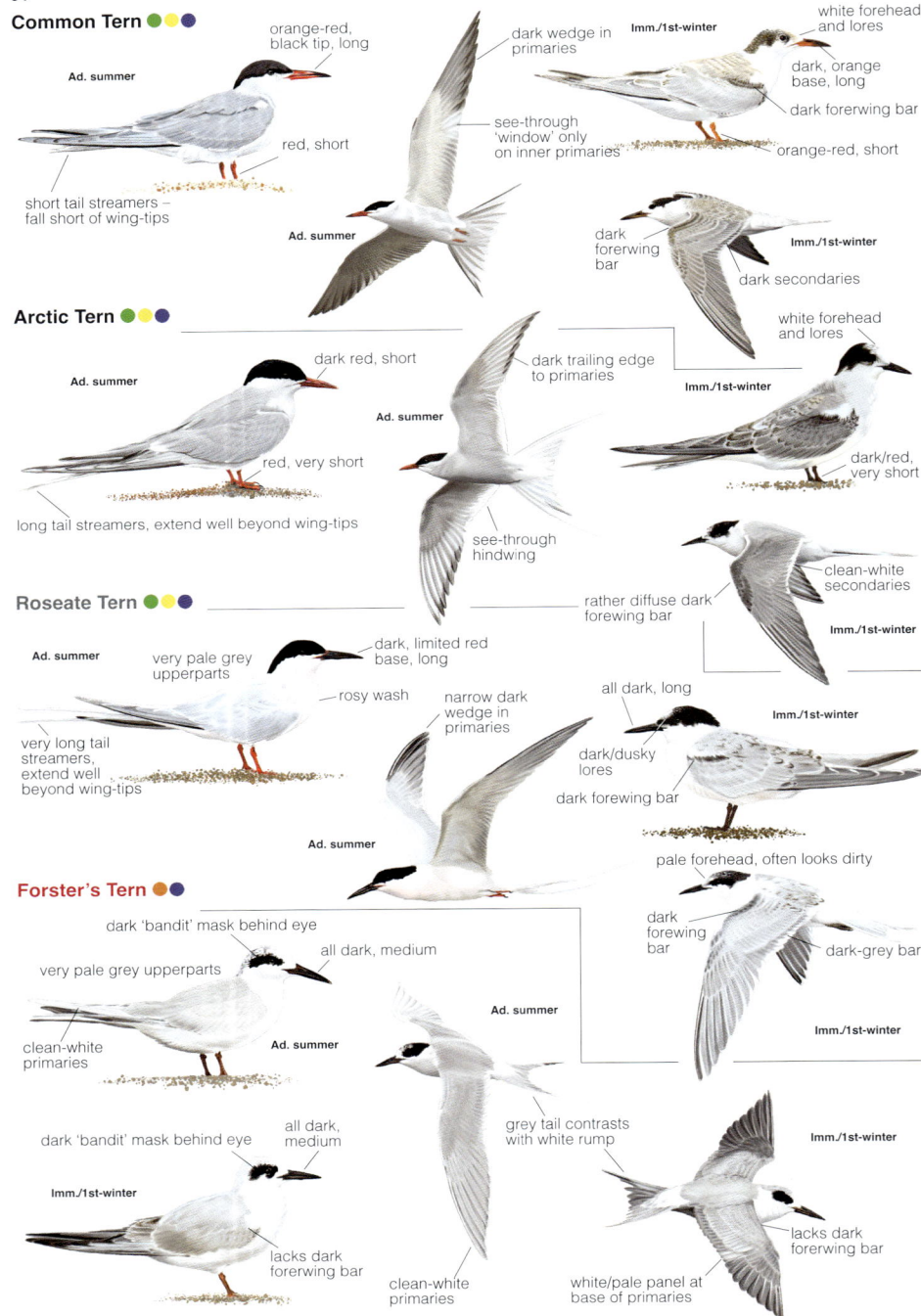

Red-billed Tropicbird 🟢🟡

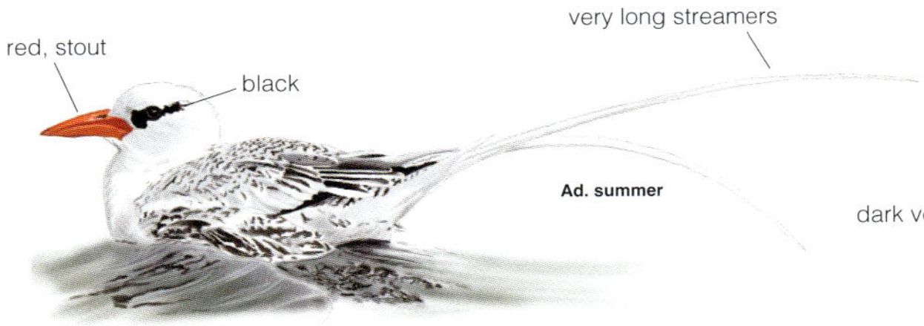

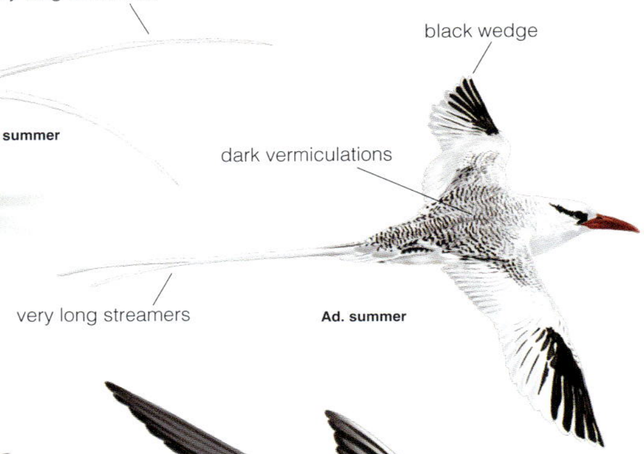

red, stout · *black* · *very long streamers* · **Ad. summer**

black wedge · *dark vermiculations* · *very long streamers* · **Ad. summer**

Aleutian Tern 🟢

all dark · *black* · *smoke grey upperparts* · *white leading edge to wing* · **Ad. summer**

white leading edge to wing · *white forehead reaches just behind eye* · **Ad. summer**

Bridled Tern 🟡

white forehead reaches just behind eye · *dark-grey/brown upperparts* · *variable grey 'collar' extends from nape to upper breast* · **Ad. summer**

Ad. summer · *dark-grey/brown upperparts* · *black* · *long streamers* · *variable grey 'collar'*

Sooty Tern 🟢🟡🟠

white forehead stops just in front of eye · *black* · *black/sooty-brown upperparts* · *clean white underparts* · **Ad. summer**

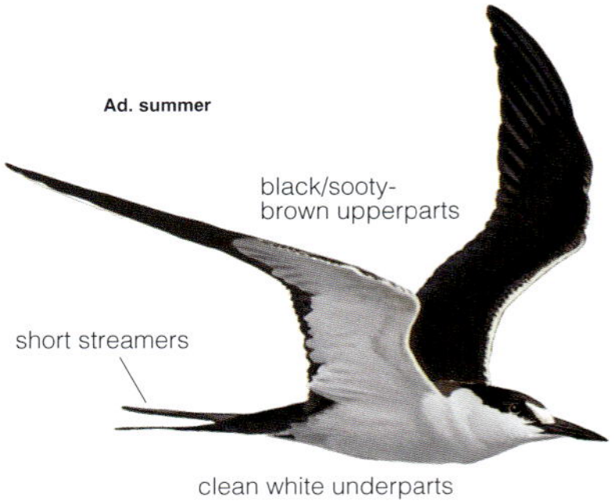

Ad. summer · *black/sooty-brown upperparts* · *short streamers* · *clean white underparts*

Little Auk

- black mask
- dark, small
- white upper eyelid
- white streaking
- winter
- winter
- white
- winter

Long-billed Murrelet

- white eye-ring
- short, slender
- pale patches
- scaly black/brown
- Imm.
- plain, dark
- Imm.

Ancient Murrelet

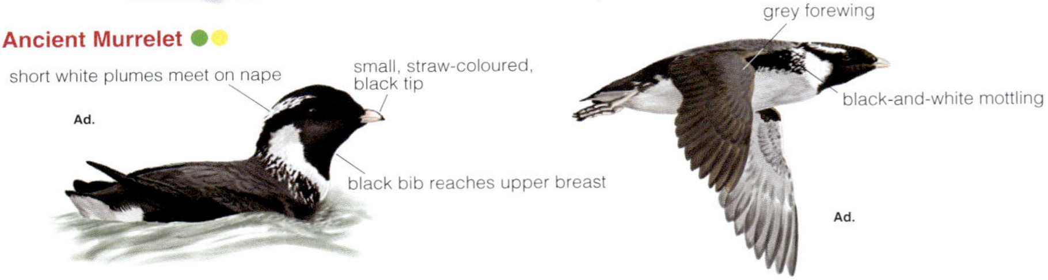

- short white plumes meet on nape
- small, straw-coloured, black tip
- Ad.
- black bib reaches upper breast
- grey forewing
- black-and-white mottling
- Ad.

Puffin

- grey 'tear-drop'
- parrot-like, red, yellow, blue
- off-white
- small, dull, red, yellow
- dusky
- summer
- plain dark wings
- summer
- all-black wings
- orange
- yellow
- winter

Tufted Puffin

- long, yellow droopy tufts
- white forehead
- Ad. summer
- large, parrot-like, two-toned orange and grey
- orange/red
- plain dark wings
- Ad. summer

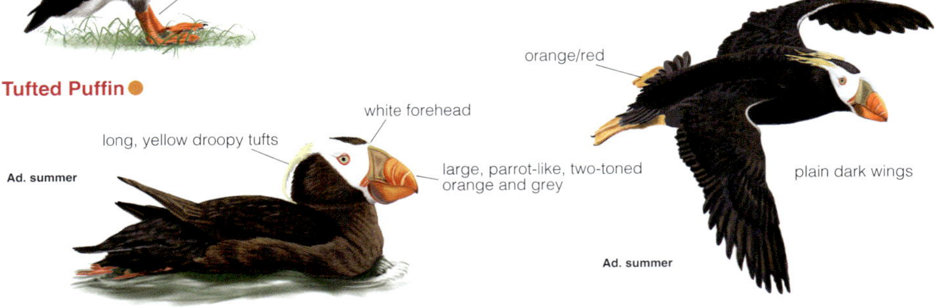

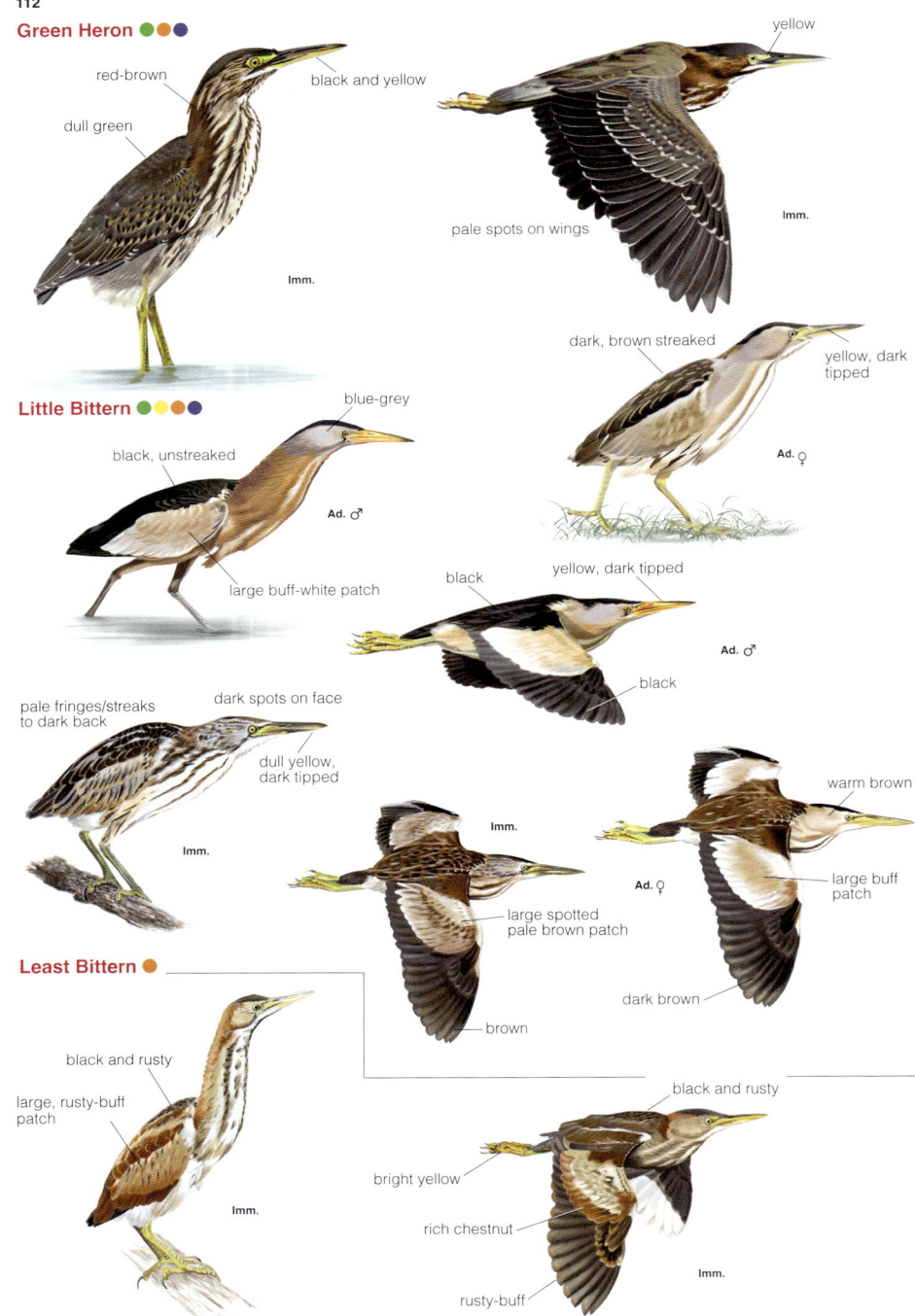

114

Great White Egret 🟢🟡🟠🟣

- long sinuous neck
- yellow
- non-breeding
- mostly black
- breeding
- long aigrettes
- yellow-green thighs
- dark, long
- long aigrettes
- some red in thighs

Little Egret 🟢🟡🟠🟣

- breeding
- long head aigrettes
- grey lores
- black
- non-breeding
- black
- yellow

Cattle Egret 🟢🟡🟠🟣

- breeding
- bright orange-yellow
- orange tinge
- yellow
- non-breeding
- yellow lores
- yellow
- short
- grey-green

Snowy Egret 🟢🟠🟣

- yellow lores
- dark
- non-breeding
- dark legs with yellow-green on back of legs
- yellow

Great

leg extension ratio in flight

Little

Cattle

Snowy

Hawk Owl 🔴🔵

Tengmalm's Owl 🟢🟡🟠🟣

Little Owl 🟢🟡🟠🟣

Scop's Owl 🟢🟡🟠🟣

127

Eastern Phoebe 🟢

- dark
- all dark
- grey smudge
- weak wingbars
- 1st-winter
- long, dark

Yellow-bellied Flycatcher 🟠

- bold eye-ring
- orange
- yellow wash
- 1st-winter
- strong buff-white wingbars
- medium primary projection

Acadian Flycatcher 🟠

- narrow, pale eye-ring
- yellow-orange, broad base
- narrow whitish-buff wingbars
- long primary projection
- yellow wash
- 1st-winter

Alder Flycatcher 🟠

- narrow, pale eye-ring
- short
- olive-green upperparts
- broad buff-white wingbars
- 1st-winter
- medium primary projection

Great Crested Flycatcher 🟠

- broad white
- pale base
- dark grey
- yellow
- rufous
- 1st-winter

Eastern Kingbird 🟠

- black
- 1st-winter
- dark grey upperparts
- pale grey band
- black
- white tip

Brown Shrike 🟢🟠

- plain brown
- heavy, stubby
- short
- plain peach
- short outer tail feathers
- **Ad. ♂**

- heavy, stubby
- plain brown
- light barring
- short
- short outer tail feathers
- **Ad. ♀**

- dark lores
- heavy, stubby
- buff
- short
- short outer tail feathers
- **1st-winter**

Red-backed Shrike 🟢🟡🟠

- blue-grey
- chestnut
- light rosy-pink underparts
- long
- white flashes
- **Ad. ♂**

- grey
- pale lores
- red-brown
- barred
- long
- **Ad. ♀**

- barred pale lores
- strongly barred
- white
- **1st-winter**

Daurian Shrike 🟠

- bright, red-brown
- sandy-grey-brown
- white primary patch
- buff wash
- long
- bright red-brown
- **Ad. ♂**

- pale lores
- sandy-grey-brown
- long
- buff wash
- **Ad. ♀**

- pale lores
- pale-sandy-brown
- long
- red-brown
- light buff wash
- **1st-winter**

Turkestan Shrike 🟠

- strong supercilium
- sandy-brown
- mostly white
- bright red-brown
- white primary patch
- white
- **Ad. ♂**

- grey-brown upperparts
- mostly white
- long
- red-brown
- **Ad. ♀**

- strong face mask
- grey-brown upperparts
- moderate/strong barring
- long
- red-brown
- **1st-winter**

Long-tailed Shrike 🟠

- blue-grey
- dark lores
- mostly white
- peach underparts
- orange-rufous
- long, graduated
- **1st-winter**

Radde's Warbler 🟠

- olive-brown upperparts
- strong, yellow-buff in front of eye, white behind
- apricot
- yellow-buff underparts
- stout, strong, pale

Dusky Warbler 🟠🔵

- strong, white in front of eye, buff behind
- dark-olive-brown upperparts
- slender
- grey-brown underparts
- thin, pink-brown

Sulphur-bellied Warbler 🟡

- strong, lemon in front of eye, white behind
- cold grey-brown upperparts
- long, slender
- pale sulphur-yellow underparts with oily grey-brown flanks and sides to breast
- dull red-brown

Hume's Warbler 🟠🔵

- strong, buff-white in front of eye, white behind
- grey-olive upperparts
- two buff-white wingbars
- dirty-white underparts
- call: 'ti-sip'

Yellow-browed Warbler 🟠🔵

- bright olive-green upperparts
- yellow-white
- two yellow-white wingbars
- silky white
- call: 'see-wee'

Pallas's Warbler 🟠🔵

- grey-green upperparts
- lemon rump
- yellow
- long, strong, yellow-white
- two yellow-white wingbars

Blyth's Reed Warbler 🟢🟡🟠

- cool brown upperparts
- bulging, pale
- dark tip to lower mandible
- 60 | 40
- short wings
- dull grey-brown

Reed Warbler 🟢🟡🟠

- dark centres
- indistinct supercilium
- long, slender
- rust-brown
- 50 | 50
- medium wings

Marsh Warbler 🟢🟡🟠

- olive-brown/grey upperparts
- slightly bulging
- short, stout
- 40 / 60
- long wings
- white tips

Paddyfield Warbler 🟢🟡🟠

- prominent, long, white
- short, dark
- black eye-stripe
- 65 / 35
- short wings
- rust brown

Cetti's Warbler 🟢🟡🟠🔵

- white tips
- 70 | 30
- short wings
- rusty/brown upperparts
- narrow, white
- grey-white

Blackcap 🟡🟠🔵

Garden Warbler 🟢🟠

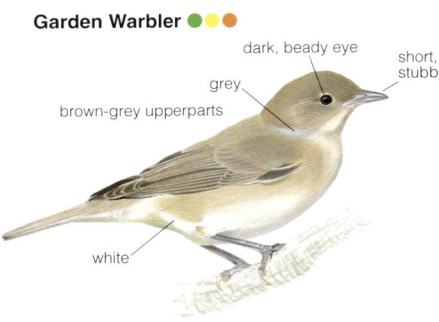

Barred Warbler 🟢🟠

Lesser Whitethroat 🟢🟠🔵

Western Orphean Warbler 🟢🟠

Eastern Orphean Warbler 🟠

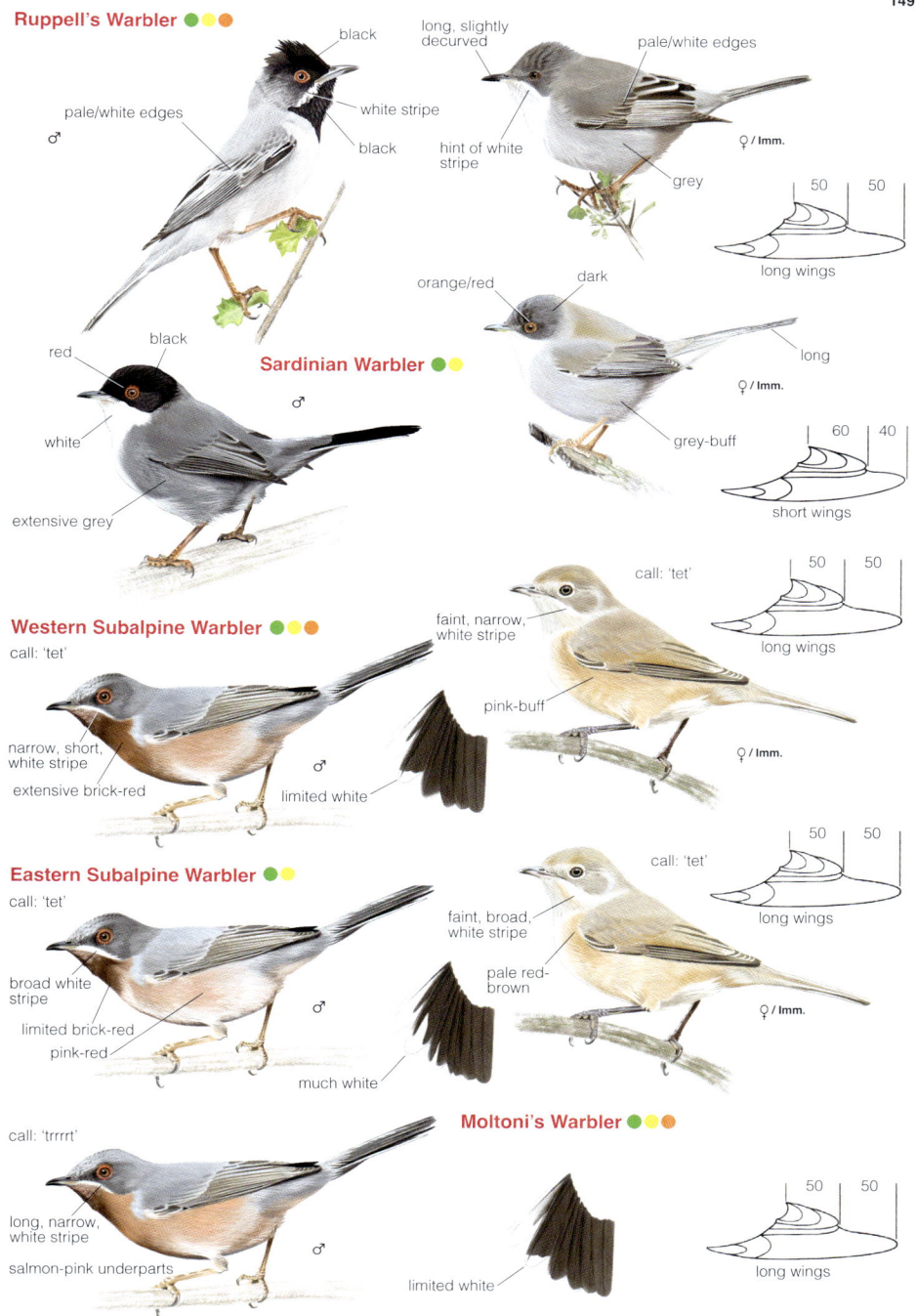

Wren 🟢🟡🟠🔵

- long, pale brown/white
- brown upperparts
- short, cocked
- heavily barred
- heavily barred, pale brown

Red-breasted Nuthatch 🟠🔵

- dark/black
- bold white
- broad, black
- blue-grey upperparts
- rusty-orange-brown underparts

Imm. ♂

Nuthatch 🟢🟡🟠🔵

- blue-grey upperparts
- long, black
- white spots
- deep rusty-brown
- rusty-buff underparts

♂

- blue-grey upperparts
- long, black
- white spots
- rusty-buff
- pale-orange-buff underparts

♀ / Imm.

Treecreeper 🟢🟡🟠🔵

- white underparts
- white spot
- long hind claw
- large step
- rounded white spots

Short-toed Treecreeper 🟠🔵

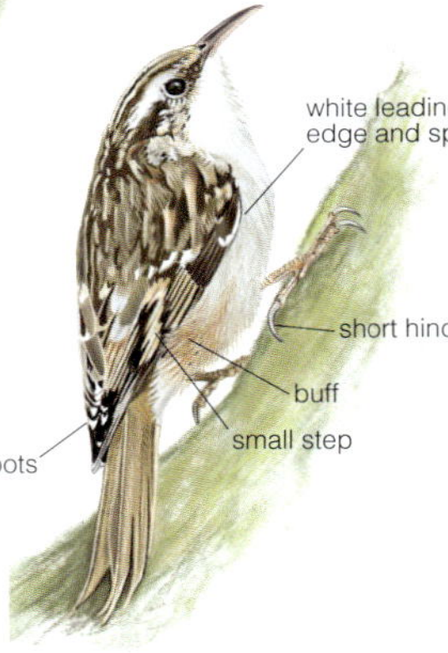

- white leading edge and spot
- short hind claw
- buff
- small step
- diamond white spots

Wallcreeper 🟢🟠🔵

- grey
- long, decurved
- deep red
- white spots
- pale tips

155

Varied Thrush 🟠

- orange
- black
- two orange wingbars
- black
- orange

Wood Thrush 🟢🟠🟣

- broad, white
- dark streaked
- warm-brown upperparts
- bold black spots
- pale pink

Imm./1st-winter

Swainson's Thrush 🟡🟠

- bold, pale
- buff lores
- grey-brown upperparts
- bold, dark spots
- grey wash

Imm./1st-winter

Hermit Thrush 🟠

- olive-brown
- narrow, dark lateral spots
- broad, dark spots
- chestnut
- buff

Imm./1st-winter

Grey-cheeked Thrush 🟠🟣

- olive-grey upperparts
- white throat
- dark/black lateral stripe
- dark spotted
- pale grey

Imm./1st-winter

Veery 🟢🟠🟣

- warm-brown upperparts
- grey
- diffuse grey/brown spots
- white underparts
- red-brown

Imm./1st-winter

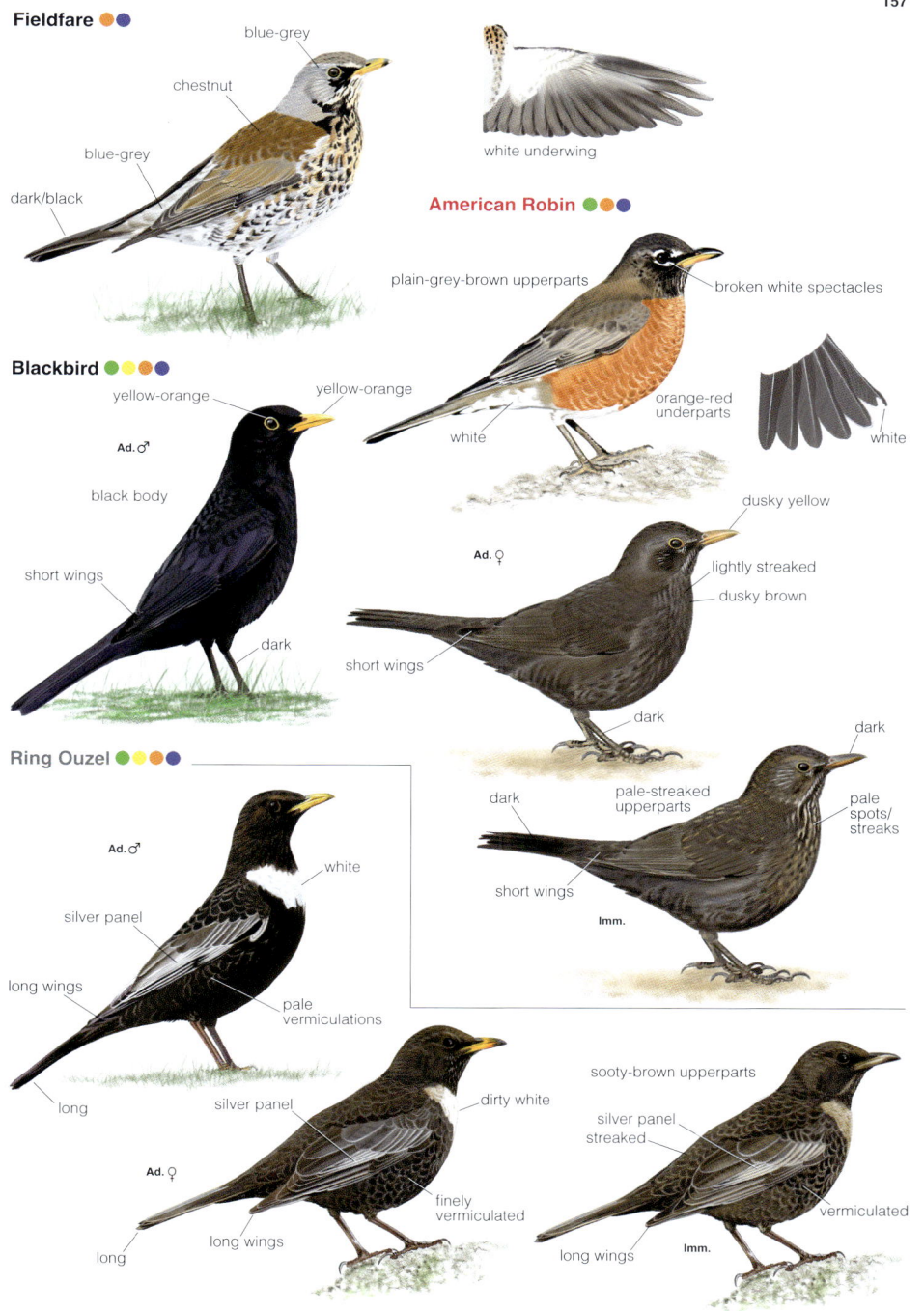

House Sparrow 🟢🟡🟠🟣

- small white spot
- grey
- small black bib
- Ad. winter ♂
- grey
- black
- Ad. summer ♂
- plain grey

- grey-brown
- buff
- dark-grey-brown stripe
- dusky grey
- ♀ / Imm.
- dusky-grey underparts

Spanish Sparrow 🟡

- broken, narrow supercilium
- chestnut
- white
- black
- ♂
- black spots

Tree Sparrow 🟢🟡🟠🟣

- warm brown
- white collar
- small, neat black bib
- black spot
- pale-grey-brown underparts

Rock Sparrow 🟢🟡🟠🟣

- broad, pale stripe
- broad, pale
- heavy, conical
- yellow-buff
- ♂
- white spots in tail

Alpine Accentor 🟡🟠🟣

- grey
- white spots
- yellow
- black-and-white spots
- red-brown, streaked

Dunnock 🟢🟡🟠🟣

- dark, slender
- red-brown eye
- grey
- warm brown, streaked
- Ad.
- dark brown, streaked

Siberian Accentor 🟠

- dark/black
- orange-buff
- dark/black
- rich buff
- lightly streaked warm brown

- slender
- grey-brown
- lightly streaked
- heavily streaked body
- pink
- Juv.

Richard's Pipit 🟢🟠🟣

- pale lores
- restricted white
- 1st-winter
- call: 'tshreep'
- moulted adult-type median covert(s) with narrow, pointed dark centres
- long hind claw

Blyth's Pipit 🟠🟣

- pale lores
- call: 'sleeoo'
- much white
- 1st-winter
- moulted adult-type median covert(s) with broad dark centres
- short hind claw

Tawny Pipit 🟢🟠

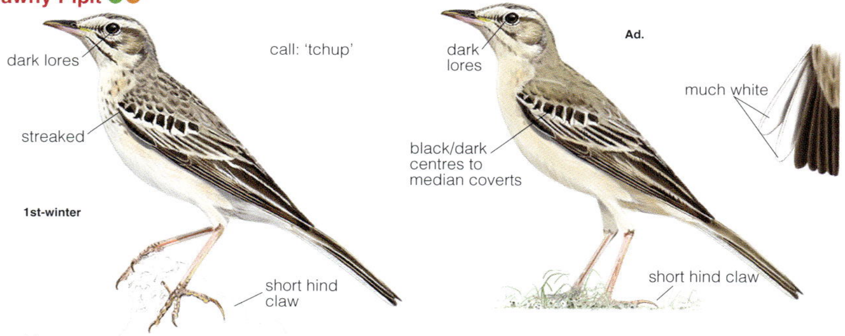

- dark lores
- streaked
- call: 'tchup'
- 1st-winter
- short hind claw
- Ad.
- dark lores
- black/dark centres to median coverts
- much white
- short hind claw

Water Pipit 🟢🟠🟣

- blue-grey
- white
- dark lores
- pink
- Ad. summer
- warm brown
- white
- dark lores
- Ad. winter
- white underparts
- well-streaked
- warm brown

Rock Pipit 🟢🟡🟠🟣 / 🟢🟠🟣

- dark lores
- diffusely streaked
- light/strong streaking
- concolourous
- blue-grey
- white
- strongly streaked, pink-buff
- grey-brown
- grey
- *littoralis*
- Ad. summer

176

Snow Bunting 🟢🟡🟠🟣

Ad. summer ♂
- clean white
- dark
- black
- large white panel
- black

♀ / Imm.
- pale buff-brown
- buff
- brown, streaked
- small white panel
- buff

Ad. winter ♂
- buff
- pale buff
- pale
- sandy-brown, streaked
- large white panel

Lapland Bunting 🟢🟠🟣

Ad. ♂
- pale cream 'C'
- chestnut
- pale
- black
- white

♀ / Imm.
- pale
- dark/black
- chestnut
- dark gorget
- two narrow white wingbars

Little Bunting 🟢🟠🟣
- black
- brown
- white eye-ring
- chestnut
- white, lightly streaked underparts

Red-headed Bunting 🟠

Imm./1st winter
- prominent streaks
- plain face
- streaked
- grey
- green-yellow

Black-headed Bunting 🟢🟠

Ad. ♂
- black
- long, heavy
- red-brown
- bright red-brown
- bright yellow underparts

♀ / Imm.
- long, heavy
- two narrow white wingbars
- bright red-brown
- yellow
- pale-yellow wash to underparts

Lark Sparrow 🟢🟡

1st-winter

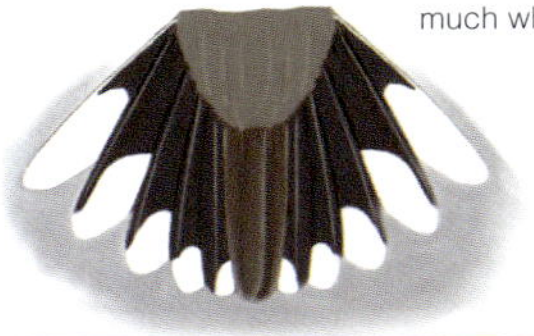

rufous stripes
broad, white
white stripe
dark spot
much white in tail
white corners

White-crowned Sparrow 🟢🟠🟣

rufous-brown stripes
broad buff-grey stripe
1st-winter
grey-brown
buff-brown
all dark

black stripes
broad, white stripe
pink/orange
Ad.
grey
all dark

White-throated Sparrow 🟢🟡🟣

narrow buff-white stripe
buff-white supercilium
Imm.
white with black border
dark streaked
all dark

yellow and white supercilium
narrow white stripe
Ad.
white with black border
grey
all dark

Savannah Sparrow 🟢🟡

yellow and off-white supercilium
narrow white stripe
strong streaks
1st-winter
strong brown streaks
all dark, notched

Bobolink 🟠

narrow yellow-buff stripe
black stripes
pale buff lores
1st-winter
yellow-buff underparts
spikey pointed tips

Song Sparrow 🟢🟠

pale-grey-white supercilium
brown, lightly streaked
strong brown streaks
dark, rounded
1st-winter
strong streaks

Booted Eagle

Abbreviations and symbols

Ad. – adult bird (fully mature adult plumage, repeated year after year).

Imm. – any plumage, other than adult, which generally refers to a young bird before reaching full adult plumage.

Juv. – juvenile (fledged young bird in its first plumage of true feathers).

1st-winter – age category referring to the plumage that replaces juvenile feathers.

1st-summer – age category referring to the plumage at roughly one year of age, and usually involves a partial moult in late winter/early spring.

2nd-winter – age category referring to the plumage normally following a complete (or partial) moult from 1st-summer plumage.

2nd-summer – age category referring to the plumage at roughly two years of age, and usually involves a partial moult in late winter/early spring.

P – primary feather e.g. p1 = innermost primary.

S – secondary feather e.g. s1 = outermost secondary.

♂ – male

♀ – female

Glossary

Aigrettes – ornamental feathers found on the back and head of heron species in the breeding season.

Alula (Bastard-wing) – group of feathers (usually three to four but can be up to seven) on the upper leading edge of the wing, just distal to the carpal joint.

Axilliaries – the 'armpits' of a bird, only revealed during wing stretching or in flight.

Bill – the beak (see mandible).

Boa – area of feathering on the lower hind-neck.

'Braces' – lines of pale/bright colour on the upper back.

Cape – extensive feathering found primarily on the neck of male pheasants.

Carpal – point at which the wing folds.

Cere – an area of bare skin on the base of the upper mandible found in some groups, e.g. raptors.

Culmen – ridge formed along the top of the bill from the forehead to the tip of the bill.

Cutting edge – edges of the upper and lower mandibles (bill) that meet to create a cutting surface.

Eye-ring – area of feathering, often pale, around the eye. Also known as 'spectacles'.

Eye-stripe – dark stripe running through the eye from the base of the bill to rear ear-coverts.

Flight feathers – the long feathers in the wing (primaries, secondaries and tertials), and tail.

Gape – the fleshy edges or base of the bill, particularly prominent in very young birds.

Gonys – the ridge towards the end of the lower mandible at the junction of the two joined halves.

Gonydeal angle – is the point where the lower mandible tips upwards near the tip (most prominent in gulls).

Grin patch – open area found in the bill of Snow Geese.

Gular pouch – an area of bare skin in the lower bill/throat area that can be extended to accommodate large items of food.

Loral spot – pale spot found in the lores (see below).

Lores – area of feathering between the eye and the bill.

Malar – stripe, usually dark, along the side of the throat, below the submoustachial stripe.

Mandible – the beak of a bird, consisting of upper and lower mandible.

Mantle – upper back, below the neck.

'Mirror' – small white patch or spot on an otherwise black feather and just inside the tip of the longest primaries (commonly seen in gulls).

Morph – plumage variant within a species and which is not geographically defined (cf. *subspecies*).

Moustachial stripe – stripe, often dark, running from base of the lower mandible along the lower edge of the cheek.

Nail – horny plate-like feature, shaped like a shield, found at the tip of the upper bill of wildfowl.

Orbital ring – unfeathered area of skin around the eye, which usually forms a distinctive ring around the eye (not to be confused with eye-ring – see above).

Primaries – main flight feathers that form the outer part of the wing, or 'fingers'.

Primary projection (primary/tertial ratio) – proportion of the primary feathers that are visible beyond the tertials on the folded wing.

Rump – area of feathering on the upperside between the back and the tail.

Scapulars – feathers between the shoulder and the mantle.

Secondaries – inner flight feathers that form the trailing edge of the inner wing.

Shield – area of bare skin found above the base of the bill. e.g. Coot and Moorhen.

Speculum – area of glossy coloured feathering found on the upper secondaries, most often on dabbling ducks.

Sub-moustachial stripe – most often a pale stripe between the moustachial stripe and malar.

Supercilium – stripe (usually pale) running horizontally above the eye.

Tarsus – the leg of a bird.

Tertials – the three feathers that cloak the bases of the flight feathers. These are usually fairly triangular in shape.

'Window' – area of 'see-through' feathering in the wing that allows some light through. Can be seen in some gulls, terns and raptors, in flight from below.

Wing – inner wing, hind and fore.

Wingbar(s) – contrastingly coloured pale tips, often white, to the wing-coverts forming a bar.

Wing panel – an area of contrasting colour, usually rectangular, within the wing.

References/further reading

Baker, J.K. (1993) *Identification of European Non-Passerines* 2nd ed. BTO Books, Thetford.

Birdguides. birdguides.com

Birding World (1988-2012) Ornithological Journal Cley-next-the-sea, Holt.

Blasco Zumeta, J. & Heinze, G.M. (2016) *Identification atlas of Aragon's birds.* www.javierblasco.arrakis.es/indexE.htm

British Ornithologists' Union. bou.org.uk

British Birds (1907-) Ornithological Journal (12 issues/year) London.

British Birds. britishbirds.co.uk

Campbell, B & Lack, E. (1985) *A Dictionary of Birds* T & A D Poyser, London.

Chandler, R.J. (1989) *North Atlantic Shorebirds* Macmillan Field Guides.

Clement, P. & Hathway, R. (2000) *Thrushes* Helm, London.

Cocker, M. & Mabey, R. (2005) *Birds Britannica* Chatto & Windus.

Cottridge, D. & Vinicombe, K. (1996) Rare Birds in Britain & Ireland. A photographic record. HarperCollins, London.

Cramp, S. Simmons, K. E. L. & Perrins, C. M. (ed) (1977-94) T*he Birds of the Western Palearctic.* Vol. 1-9 Oxford University Press, Oxford.

Flood, B. & Fisher, A. (2016) *North Atlantic Seabirds* Flood and Fisher Multimedia ID Guides.

Flood, R. L. Hudson, N.& Thomas, B. (2007) *Essential Guide to Birds of the Isles of Scilly* Nigel Hudson.

Forsman, D. (1999) *The Raptors of Europe and The Middle East* T & A D Poyser, London.

Gibbs, D., Barnes, E., & Cox, J. (2001) *Pigeons and Doves* Helm, London.

Grant, P.J. (1982) *Gulls: A Guide to Identification* T & A D Poyser.

Harrison, P. (1985) *Seabirds* Helm, London.

Hayman, P. Marchant, J. & Prater, T. (1991) *Shorebirds: An Identification Guide to the Waders of the World* Helm, London.

Jenni, L. & Winkler, R. (1994) *Moult and Ageing of European Passerines* Academic Press, London.

Lee, C. & Birch, A. (2023) *Field Guide to North American Flycatchers* Princeton.

Macauley library The Cornell Lab of Ornithology macauleylibrary.org

Madge, S. & Burn, H. (2010) *Wildfowl* Helm, London.

Madge, S. McGowan, P. Arlott, N. Budden, R. Cole, D. Cox, J. D'Silva, C. Franklin, K. & Mead, D. (2002) *Pheasants, Partridges and Grouse* Helm, London.

Malling, K. & Larsson, H. (1995) *Terns of Europe and North America* Helm, London.

Malling, K. & Larsson, H. (1997) *Skuas And Jaegars* Pica, Mountfield.

Mitchell, D & Young, S. (1997) Rare Birds of Britain and Europe New Holland, London.

Norevik, G. Hellstrom, M. Liu, D. & Petersson, B. (2020) *Ageing & Sexing of Migratory East Asian Passerines* Avium Forlag.

Olsen, M.K. (2004) *Gulls of Europe, Asia and North America* Helm, London.

Porter, R. F. Willis, I. Christensen, S. & Pors Neilsen, B. (1981) *Flight Identification of European Raptors* T & A D Poyser, London.

Sibley, D. A. (2000) *The North American Bird Guide* Pica Press, East Sussex.

Shirihai, H. & Svensson, L. (2018) *Birds of the Western Palearctic Volume 1 Passerines: Larks to Warblers* Helm, London.

Shirihai, H. & Svensson, L. (2018) *Birds of the Western Palearctic Volume 2 Passerines: Flycatchers to Buntings* Helm, London.

Shirihai, H. Gargallo, G. & Helbig, A, J. Harris, A. Cotteridge, D. (2001) *Sylvia Warblers* Helm, London.

Shirihai, H. Christie, D. & Harris, A. (1996) *Birders Guide to European and Middle Eastern Birds* Macmillan, London.

Stephenson, T. & Whittle, S. (2013) *The Warbler Guide* Princeton University Press, Oxfordshire.

Sterry, P. & Stancliffe, P. (2015) *Collins BTO Guide To British Birds* Collins.

Sterry, P. & Stancliffe, P. (2015) *Collins BTO Guide To Rare British Birds* Collins.

Svennson, L. Mullarney, K. & Zetterstrom, D. (2023) *Collins Bird Guide, 1st, 2nd & third editions* Collins.

Svensson, L. (1992) *Identification Guide to European Passerines* 4th ed. Stockholm.

Vinicombe, K. Harris, A. & Tucker, L. (1989) *Bird Identification* Macmillan, London.

xeno-canto Sharing wildlife sounds from around the World xeno-canto.org

Paul Stancliffe

Paul currently lives with his wife in central Portugal, close to the Spanish border. He has worked for the British Trust for Ornithology, published two photographic identification guides to British Birds with Collins, lived on the Isles of Scilly, where he helped form the Isles of Scilly Bird Group. He has been a member of several local rarities panels, has been a member of the Bird Observatories Council and is on the advisory panel of the grant awarding charity, Birds on the Brink, supporting bird conservation around the world.

Jeff Baker

After 48 years at the BTO (British Trust for Ornithology) retirement allowed Jeff the time to devote his attention to illustrating this book. His past publications *Warblers of Europe, Asia and North Africa* (Helm Identification Guides) and *Identification of European Non-Passerines* (BTO Books) were key reference sources that helped to bring about this latest offering. He has been a keen birder since the age of 12 and a licensed bird ringer since 1970. He has travelled extensively, usually in search of warblers (personal), and also helped train others in bird ringing (professional). He would like to properly retire, but that seems unlikely!

Index

Acanthis flammea, 172
Accentor, Alpine, *Prunella collaris*, 166
    Siberian, *Prunella montanella*, 166
Accipiter nisus, 117
Acrocephalus agricola, 145
    arundinaceus, 144
    dumetorum, 145
    paludicola, 144
    palustris, 145
    schoenobaenus, 144
    scirpaceus, 145
Actitis hypoleucos, 83
    macularius, 83
Aegolius funereus, 120
Agelaius phoeniceus, 181
Aix, galericulata, 37
Alauda arvensis, 134
    leucoptera, 135
Albatross, Black-browed, *Thalassarche melanophris*, 104
    Yellow-nosed, *Thalassarche chlororhynchos*, 104
Alca torda, 101
Alcedo atthis, 122
Alectoris rufa, 30
Alle alle, 102
Anarhynchus alexandrinus, 71
    atrifrons, 71
    asiaticus, 71
    leschenaultii, 71
    mongolus, 71
Anas, acuta, 41
    crecca, 38
    platyrhynchos, 40
    rubripes, 41
Anser, anser, 34
    albifrons, 35
    brachyrhynchus, 34
    carolinensis, 38
    caerulescens, 33
    erythropus, 35
    fabalis, 34
    rossii, 33
Anthus campestris, 168
    cervinus, 169
    godlewskii, 168
    gustavi, 169
    hodgsoni, 169
    petrosus, 168
    pratensis, 169
    richardi, 168
    rubescens, 169
    spinoletta, 168
    trivialis, 169
Antigone canadensis, 65
Apus, affinis, 57
    apus, 56
    caffer, 57
    pacificus, 57
    pallidus, 56
Aquila chrysaetos, 115
    pennata, 183
Ardea alba, 114
    cinerea, 111
    herodias, 111
    ibis, 114
    purpurea, 111
Ardenna gravis, 107
    grisea, 107
Ardeola bacchus, 113
    ralloides, 113
Arundinax aedon, 144
Asio otus, 121
    flammeus, 121
Astur gentilis, 117
Athene noctua, 120
Auk, Little, *Alle alle*, 102
Avocet, *Recurvirostra avosetta*, 67
Aythya affinis, 45
    americana, 44
    collaris, 45
    ferina, 44
    marila, 45
    fuligula, 45
    nyroca, 45
    valisineria, 44

Bartramia longicauda, 72
Bee-eater,
    Blue-cheeked, *Merops persicus*, 122
    (European), *Merops apiaster*, 122
Bittern,
    American, *B. lentiginosus*, 111
    (Eurasian), *Botaurus stellaris*, 111
    Least, *Botaurus exilis*, 112
    Little, *Botaurus minutus*, 112

Blackbird,
- Common *Turdus merula,* 157
- Red-winged, *Agelaius phoeniceus,* 181

Blackcap, *Sylvia atricapilla,* 148
Bluetail, Red-flanked, *Tarsiger cyanurus,* 160
Bluethroat, *Luscinia svecica* 160
Bobolink, *Dolichonyx oryzivorus,* 180
Bombycilla cedrorum, 153
- *garrulus,* 153

Booby, Brown, *Sula leucogaster,* 98
- Red-footed, *Sula sula,* 98

Botaurus stellaris, 111
Brachyramphus perdix, 102
Brambling, *Fringilla montifringilla,* 170
Branta bernicla, 32
- *canadensis,* 33
- *hutchinsii,* 33
- *leucopsis,* 33
- *serrirostris,* 34

Bubo scandiacus, 121
Bucanetes githagineus, 173
Bucephala, albeola, 52
- *islandica,* 52
- *strepera,* 52

Bufflehead, *Bucephala albeola,* 52
Bullfinch, *Pyrrhula pyrrhula,* 170
Bunting, Black-faced, *Emberiza spodocephala,* 179
- Black-headed, *Emberiza melanocephala,* 176
- Chestnut-eared, *Emberiza fucata,* 177
- Cirl, *Emberiza cirlus,* 178
- Chestnut, *Emberiza rutila,* 178
- Corn, *Emberiza calandra,* 179
- Cretzschmar's, *Emberiza caesia,* 179
- Indigo, *Passerina cyanea,* 174
- Lapland, *Calcarius lapponicus,* 176
- Little, *Emberiza pusilla,* 176
- Ortolan, *Emberiza hortulana,* 179
- Pallas's Reed, *Emberiza pallasi,* 177
- Pine, *Emberiza leucocephalos,* 178
- Red-headed, *Emberiza bruniceps,* 176
- Reed, *Emberiza schoeniclus,* 177
- Rock, *Emberiza cia,* 179
- Rustic, *Emberiza rustica,* 177
- Snow, *Plectrophenax nivalis,* 176
- Yellow-breasted, *Emberiza aureola,* 178
- Yellow-browed, *Emberiza chrysophrys,* 177

Burhinus oedicnemus, 67
Bustard,
- Asian Houbara, *Chlamydotis macqueenii,* 58
- Great, *Otis tarda,* 28
- Little, *Tetrax tetrax,* 58

Buteo, buteo, 116
- *lagopus,* 116
- *rufinus,* 116

Buzzard,
- Common, *Buteo buteo,* 116
- Honey, *Pernis apivorus,* 116
- Long-legged, *Buteo rufinus,* 116
- Rough-legged, *Buteo lagopus,* 116

Calandrella brachydactyla, 134
Calcarius lapponicus, 176
Calliope calliope, 160
Calidris acuminata, 75
- *alba,* 77
- *alpina,* 78
- *bairdii,* 77
- *canutus,* 74
- *falcinellus,* 77
- *ferruginea,* 74
- *fuscicollis,* 77
- *himantopus,* 74
- *maritima,* 78
- *mauri,* 78
- *melanotos,* 75
- *minuta,* 76
- *minutilla,* 76
- *pugnax,* 75
- *pusilla,* 78
- *ruficollis,* 76
- *subminuta,* 76
- *subruficollis,* 75
- *temminckii,* 76

Calonectris borealis, 107
- *diomedea,* 107

Canvasback, *Aythya valisineria*, 44
Capercaillie, *Tetrao urogallus*, 29
Caprimulgus aegyptius, 55
    europaeus, 55
    fuscescens, 155
    ruficollis, 55
Cardellina pusilla, 138
Carduelis carduelis, 171
    citrinella, 171
Carpodacus erythrinus, 173
Catbird, Grey, *Dumetella carolinensis*, 153
Catharus guttatus, 155
    minimus, 155
    ustulatus, 155
Cecropis daurica, 182
    rufula, 137
Cepphus grylle, 101
Cercotrichas galactotes, 159
Certhia brachydactyla, 152
    familiaris, 152
Cettia cetti, 145
Chaetura pelagica, 57
Chaffinch, *Fringilla coelebs*, 170
Charadrius dubius, 70
    hiaticula, 70
    semipalmatus, 70
    vociferus, 70
Chiffchaff, Common, *Phylloscopus collybita*, 141
    Iberian, *Phylloscopus ibericus*, 141
    Siberian, *Phylloscopus collybita tristis*, 141
Chlamydotis macqueenii, 58
Chlidonias hybrida, 97
    leucopterus, 97
    niger, 97
Chloris chloris, 170
Chondestes grammacus, 180
Chordeiles minor, 55
Chough, *Pyrrhocorax pyrrhocorax*, 131
Chroicocephalus genei, 87
    philadelphia, 87
    ridibundus, 87
Chrysolophus amherstiae, 31
    pictus, 31
Ciconia ciconia, 109
    nigra, 109
Cinclus cinclus, 154

Circaetus gallicus, 115
Circus aeruginosus, 118
    cyaneus, 119
    hudsonius, 119
    macrourus, 119
    pygargus, 119
Cisticola juncidis, 144
Clamator glandarius, 59
Clanga clanga, 115
Clangula hyemalis, 51
Coccothraustes coccothraustes, 173
Coccyzus americanus, 59
    erythropthalmus, 59
Coloeus monedula, 131
Columba livia, 60
    oenas, 60
    palumbus, 60
Coot,
    American, *Fulica americana*, 64
    (Eurasian), *Fulica atra*, 64
Coracias garrulus, 122
Cormorant,
    Double-crested, *Nannopterum auritum*, 110
    (Great), *Phalacrocorax carbo*, 110
Corncrake, *Crex crex*, 30
Corthylio calendula, 151
Corvus corax, 131
    corone, 131
    cornix, 131
    frugilegus, 131
Coturnix, coturnix, 30
Courser, Cream-coloured, *Cursorius cursor*, 85
Cowbird, Brown-headed, *Molothrus ater*, 181
Crake, Baillon's, *Zapornia pusilla*, 63
    Little, *Zapornia parva*, 63
    Spotted, *Porzana porzana*, 63
Crane, Common, *Grus grus*, 65
    Sandhill, *Antigone canadensis*, 65
Crex crex, 30
Crossbill,
    Parrot, *Loxia pytyopsittacus*, 175
    Scottish, *Loxia scotica*, 175
    (Red), *Loxia curvirostra*, 175
    Two-barred, *Loxia leucoptera*, 175
Crow, Carrion, *Corvus corone*, 131
    Hooded, *Corvus cornix*, 131
Cuculus canorus, 59

Cuckoo,
    Black-billed, *Coccyzus erythropthalmus,* 59
    (Common), *Cuculus canorus,* 59
    Great Spotted, *Clamator glandarius,* 59
    Yellow-billed, *Coccyzus americanus,* 59
Curlew, *Numenius arquata,* 72
Curruca cantillans, 149
    communis, 150
    conspicillata, 150
    crassirostris, 148
    curruca, 148
    curruca blythi, 148
    hortensis, 148
    iberiae, 149
    melanocephala, 149
    nana, 150
    nisoria, 148
    ruppeli, 149
    sarda, 150
    subalpina, 149
    undata, 150
Cursorius cursor, 85
Cyanistes caeruleus, 133
Cygnus columbianus, 36
    cygnus, 36
    olor, 36

Delichon urbicum, 137
Dendrocopos major, 124
Dipper, *Cinclus cinclus,* 154
Diver,
    Black-throated, *Gavia arctica,* 103
    Great Northern, *Gavia immer,* 103
    Pacific, *Gavia pacifica,* 103
    Red-throated, *Gavia stellata,* 103
    White-billed, *Gavia adamsii,* 103
Dolichonyx oryzivorus, 180
Dotterel, *Eudromias morinellus,* 72
Dove,
    Collared, *Streptopelia decaocto,* 61
    Mourning, *Zenaida macroura,* 60
    Oriental Turtle, *Streptopelia orientalis,* 61
    Rock, *Columba livia,* 60
    Stock, *Columba oenas,* 60
    Turtle, *Streptopelia turtur,* 61

Dowitcher,
    Long-billed, *Limnodromus scolopaceus,* 79
    Short-billed, *Limnodromus griseus,* 79
Dryobates minor, 123
Duck, Black, *Anas rubripes,* 41
    Baikal, *Sibirionetta formosa,* 38
    Falcated, *Mareca falcata,* 42
    Ferruginous, *Aythya nyroca,* 45
    Harlequin, *Histrionicus histrionicus,* 51
    Long-tailed, *Clangula hyemalis,* 51
    Mandarin, *Aix galericulata,* 37
    Ringed-necked, *Aythya collaris,* 45
    Ruddy, *Oxyura jamaicensis,* 44
    Tufted, *Aythya fuligula,* 45
Dumetella carolinensis, 153
Dunlin, *Calidris alpina,* 78

Eagle, Booted, *Aquila pennata,* 183
    Golden, *Aquila chrysaetos,* 115
    Short-toed, *Circaetus gallicus,* 115
    Spotted, *Clanga clanga,* 115
    White-tailed, *Haliaeetus albicilla,* 115
Egret, Cattle, *Ardea ibis,* 114
    Great White, *Ardea alba,* 114
    Little, *Egretta garzetta,* 114
    Snowy, *Egretta thula,* 114
Egretta garzetta, 114
    thula, 114
Eider, Common *Somateria mollissima,* 47
    King, *Somateria spectabilis,* 47
    Steller's, *Polysticta stelleri,* 47
Elanus caeruleus, 126
Emberiza aureola, 178
    bruniceps, 176
    caesia, 179
    calandra, 179
    chrysophrys, 177
    cia, 179
    cirlus, 178
    citrinella, 178
    fucata, 177
    hortulana, 179
    leucocephalos, 178
    melanocephala, 176
    pallasi, 177

pusilla, 176
rustica, 177
rutila, 178
schoeniclus, 177
spodocephala, 179
Empidonax alnorum, 127
flaviventris, 127
virescens, 127
Eremophila alpestris, 135
Erithacus rubecula, 159
Eudromias morinellus, 72

Falco amurensis, 125
columbarius, 124
eleonorae, 126
naumanni, 124
peregrinus, 126
rusticolus, 126
sparverius, 124
subbuteo, 125
tinnunculus, 124
vespertinus, 125
Falcon, Amur, *Falco amurensis,* 125
Eleonora's, *Falco eleonorae,* 126
Gyr, *Falco rusticolus,* 126
Red-footed, *Falco amurensis,* 125
Ficedula albicilla, 161
albicollis, 161
hypoleuca, 161
parva, 161
Fieldfare, *Turdus pilaris,* 156
Finch, Citril, *Carduelis citrinella,* 171
Trumpeter, *Bucanetes githagineus,* 173
Firecrest, *Regulus ignicapilla,* 151
Flycatcher,
Acadian, *Empidonax virescens,* 127
Alder, *Empidonax alnorum,* 127
Asian Brown, *Muscicapa dauurica,* 161
Collared, *Ficedula albicollis,* 161
Pied, *Ficedula hypoleuca,* 161
Red-breasted, *Ficedula parva,* 161
Spotted, *Muscicapa striata,* 161
Taiga, *Ficedula albicilla,* 161
Yellow-bellied, *Empidonax flaviventris,* 127
Fratercula arctica, 102
cirrhata, 102

Fregata aquila, 104
magnificens, 104
Frigatebird, Ascension, *Fregata aquila,* 104
Magnificent, *Fregata magnificens,* 104
Fringilla coelebs, 170
montifringilla, 170
Fulica americana, 64
atra, 64
Fulmar, (Northern), *Fulmarus glacialis,* 106
Fulmarus glacialis, 106

Gadwall, *Mareca strepera,* 40
Galerida cristata, 134
Gallinago delicata, 80
gallinago, 80
media, 80
Gallinula chloropus, 64
Gallinule, Allen's, *Porphyrio alleni,* 62
American Purple, *Porphyrio martinica,* 62
Gannet, *Morus bassanus,* 98
Garganey, *Spatula querquedula,* 38
Garrulus glandarius, 130
Gavia adamsii, 103
arctica, 103
immer, 103
pacifica, 103
stellata, 103
Gelochelidon nilotica, 93
Geokichla sibirica, 158
Geothlypis trichas, 139
Glareola maldivarum, 85
nordmanni, 85
pratincola, 85
Godwit,
Bar-tailed, *Limosa lapponica,* 73
Black-tailed, *Limosa limosa,* 73
Hudsonian, *Limosa haemastica,* 73
Goldcrest, *Regulus regulus,* 151
Goldeneye, (Common), *Bucephala clangula,* 52
Barrow's, *Bucephala islandica,* 52
Goldfinch, *Carduelis carduelis,* 171
Goosander, *Mergus merganser,* 53
Goose,
Barnacle, *Branta leucopsis,* 33

Brent, *Branta bernicla,* 32
Cackling, *Branta hutchinsii,* 33
Canada, *Branta canadensis,* 33
Egyptian, *Alopochen aegyptiaca,* 37
Greylag, *Anser anser,* 34
Lesser White-fronted, *Anser erythropus,* 35
Pink-footed *Anser brachyrhynchus,* 35
Ross's, *Anser rossii,* 33
Snow, *Anser anser,* 33
Taiga Bean, *Anser fabalis,* 34
Tundra Bean, *Anser serrirostris,* 34
White-fronted, *Anser albifrons,* 35
Goshawk, *Astur gentilis,* 117
Grebe, Black-necked, *Podiceps nigricollis,* 66
  Great Crested, *Podiceps cristatus,* 66
  Little, *Tachybaptus ruficollis,* 65
  Pied-billed, *Podilymbus podiceps,* 65
  Red-necked, *Podiceps grisegena,* 66
  Slavonian, *Podiceps auritus,* 66
Greenfinch, *Chloris chloris,* 170
Greenshank, *Tringa nebularia,* 84
Grosbeak, Evening, *Hesperiphona vespertina,* 173
  Pine, *Pinicola enucleator,* 173
Grouse, Black, *Lyrurus tetrix,* 29
  Red, *Lagopus scotica,* 29
  Rose-breasted, *Pheucticus ludovicianus,* 174
Grus grus, 65
Guillemot,
  Black, *Cepphus grylle,* 101
  Brünnich's, *Uria lomvia,* 101
  Common, *Uria aalge,* 101
Gull, American Herring, *Larus smithsonianus,* 92
  Audouin's, *Ichthyaetus audouinii,* 90
  Black-headed, *Chroicocephalus ridibundus,* 87
  Bonaparte's, *Chroicocephalus philadelphia,* 87
  Cape, *Larus dominicanus,* 182
  Caspian, *Larus cachinnans,* 92
  Common, *Larus canus,* 88
  Franklin's, *Leucophaeus pipixcan,* 88
  Glaucous, *Larus hyperboreus,* 91
  Glaucous-winged, *Larus glaucescens,* 91
  Great Black-backed, *Larus marinus,* 89
  Great Black-headed, *Ichthyaetus ichthyaetus,* 90
  Herring, *Larus argentatus,* 92
  Iceland, *Larus glaucoides,* 91
  Ivory, *Pagophila eburnea,* 90
  Laughing, *Leucophaeus atricilla,* 88
  Little, *Hydrocoloeus minutus,* 86
  Mediterranean, *Ichthyaetus melanocephalus,* 87
  Ring-billed, *Larus delawarensis,* 88
  Ross's, *Rhodostethia rosea,* 86
  Sabine's, *Xema sabini,* 86
  Slaty-backed, *Larus schistisagus,* 89
  Slender-billed, *Chroicocephalus genei,* 87
  Yellow-legged, *Larus michahellis,* 92
Gulosus aristotelis, 110

Haematopus ostralegus, 67
Haliaeetus albicilla, 115
Harrier, Hen, *Circus cyaneus,* 119
  Marsh, *Circus aeruginosus,* 118
  Montagu's, *Circus pygargus,* 118
  Northern, *Circus hudsonius,* 119
  Pallid, *Circus macrourus,* 119
Hawfinch, *Coccothraustes coccothraustes,* 173
Helopsaltes certhiola, 147
Heron, Chinese Pond, *Ardeola bacchus,* 113
  Great Blue, *Ardea herodias,* 111
  Green, *Butorides virescens,* 112
  Grey, *Ardea cinerea,* 111
  Night, *Nycticorax nycticorax,* 113
  Purple, *Ardea purpurea,* 111
  Squacco, *Ardeola ralloides,* 113
Hesperiphona vespertina, 173
Himantopus, himantopus, 67

Hippolais icterina, 146
    olivetorum, 146
    polyglotta, 146
Hirundapus, caudacutus, 56
Hirundo rustica, 137
Histrionicus, histrionicus, 51
Hobby, *Falco subbuteo,* 125
Hoopoe, *Upupa epops,* 122
Hydrobates castro, 182
    leucorhous, 105
    monorhis, 105
    pelagicus, 105
Hydrocoloeus minutus, 86
Hydroprogne caspia, 95
Hylocichla mustelina, 155

Ibis, Glossy, *Plegadis falcinellus,* 110
Ichthyaetus audouinii, 90
    ichthyaetus, 90
    melanocephalus, 87
Icterus galbula, 181
Iduna caligata, 146
    opaca, 146
    pallida, 146
    rama, 146
Irania gutturalis, 160
Ixoreus naevius, 155

Jackdaw, *Coloeus monedula,* 131
Jay, *Garrulus glandarius,* 130
Junco, Dark-eyed, *Junco hyemalis,* 181
Junco hyemalis, 181
Jynx torquilla, 123

Kestrel,
    American, *Falco sparverius,* 124
    (Common), *Falco tinnunculus,* 124
    Lesser *Falco naumanni,* 124
Killdeer, *Charadrius vociferus,* 70
Kingbird, Eastern, *Tyrannus tyrannus,* 127
Kingfisher,
    Belted, *Megaceryle alcyon,* 122
    (Common), *Alcedo atthis,* 122
Kinglet, Ruby-crowned, *Corthylio calendula,* 151
Kite, Black, *Milvus migrans,* 118
    Black-winged, *Elanus caeruleus,* 126
    Red, *Milvus milvus,* 118
Kittiwake, *Rissa tridactyla,* 86
Knot, *Calidris canutus,* 74

Lagopus muta, 29
    scotica, 29
Lanius collurio, 128
    cristatus, 128
    excubitor, 129
    isabellinus, 128
    minor, 129
    nubicus, 129
    phoenicuroides, 128
    schach, 128
    senator, 129
Lapwing, (Northern), *Vanellus vanellus,* 68
    Grey-headed, *Vanellus cinereus,* 68
    Sociable, *Vanellus gregarius,* 68
    White-tailed, *Vanellus leucurus,* 68
Lark, Bimaculated, *Melanocorypha bimaculata,* 135
    Black, *Melanocorypha yeltoniensis,* 135
    Calandra, *Melanocorypha calandra,* 135
    Crested, *Galerida cristata,* 134
    Shore, *Eremophila alpestris,* 135
    Short-toed, *Calandrella brachydactyla,* 134
    White-winged, *Alauda leucoptera,* 135
Larus argentatus, 92
    cachinnans, 92
    canus, 88
    delawarensis, 88
    dominicanus, 182
    fuscus, 89
    glaucescens, 91
    glaucoides, 91
    hyperboreus, 91
    marinus, 89
    michahellis, 92
    schistisagus, 89
    smithsonianus, 92
Larvivora cyane, 159
    sibilans, 159
Leiothlypis peregrina, 139
Leucophaeus atricilla, 88

pipixcan, 88
Limnodromus, griseus, 79
    scolopaceus, 79
Limosa haemastica, 73
    lapponica, 73
    limosa, 73
Linaria cannabina, 172
    flavirostris, 172
Linnet, *Linaria cannabina,* 172
Locustella fluviatilis, 147
    lanceolata, 147
    luscinioides, 147
    naevia, 147
Lophodytes cucullatus, 53
Lophophanes cristatus, 132
Loxia curvirostra, 175
    leucoptera, 175
    pytyopsittacus, 175
    scotica, 175
Lullula arborea, 134
Luscinia luscinia, 159
    megarhynchos, 159
    svecica, 160
Lyrurus tetrix, 29

Magpie, *Pica pica,* 130
Mallard, *Anas platyrhynchos,* 40
Mareca, americana, 42
    falcata, 42
    penelope, 42
    strepera, 40
Martin, Crag, *Ptyonoprogne rupestris,* 136
    House, *Delichon urbicum,* 137
    Purple, *Progne subis,* 136
    Sand, *Riparia riparia,* 136
Megaceryle alcyon, 122
Melanitta, americana, 49
    deglandi, 49
    fusca, 49
    nigra, 49
    perspicillata, 49
Melanocorypha bimaculata, 135
    calandra, 135
    yeltoniensis, 135
Melospiza melodia, 180
Merganser, Hooded, *Lophodytes cucullatus,* 53
    Red-breasted, *Mergus serrator,* 53
Mergellus albellus, 53

Mergus merganser, 53
    serrator, 53
Merlin, *Falco columbarius,* 124
Merops apiaster, 122
    persicus, 122
Milvus migrans, 118
    milvus, 118
Mimus polyglottos, 153
Mniotilta varia, 140
Mockingbird, Northern, *Mimus polyglottos,* 153
Molothrus ater, 181
Motacilla alba, 167
    alba yarrellii, 167
    cinerea, 167
    citreola, 167
    flava, 167
    tschutschensis, 167
Monticola saxatilis, 162
    solitarius, 162
Moorhen, *Gallinula chloropus,* 64
Morus bassanus, 98
Murrelet, Ancient, *Synthliboramphus antiquus,* 102
    Long-billed, *Brachyramphus perdix,* 102
Muscicapa dauurica, 161
    striata, 161

Nannopterum auritum, 110
Needletail, White-throated, *Hirundapus caudacutus,* 56
Neophron percnopterus, 115
Netta, rufina, 44
Nighthawk (Common), *Chordeiles minor,* 55
Nightingale, (Common), *Luscinia megarhynchos,* 159
    Thrush, *Luscinia luscinia,* 159
Nightjar, Egyptian, *Caprimulgus aegyptius,* 55
    (European), *Caprimulgus europaeus,* 55
    Red-necked, *Caprimulgus ruficollis,* 55
Nucifraga caryocatactes, 130
Numenius arquata, 72
    hudsonicus, 72
    minutus, 72
    phaeopus, 72

Nutcracker, *Nucifraga caryocatactes,* 130
Nuthatch, (Eurasian), *Sitta europaea,* 152
   Red-breasted, *Sitta canadensis,* 152
Nycticorax nycticorax, 113

Oenanthe deserti, 164
  hispanica, 165
  hispanica hispanica, 165
  hispanica melanoleuca , 165
  isabellina, 164
  leucopyga, 165
  oenanthe, 164
  pleschanka, 165
Onychoprion aleuticus, 96
  anaethetus, 96
  fuscatus, 96
Oriole,
  Baltimore, *Icterus galbula,* 181
  Golden, *Oriolus oriolus,* 130
  Indian, *Oriolus kundoo,* 130
Oriolus kundoo, 130
  oriolus, 130
Osprey, *Pandion haliaetus,* 115
Otis tarda, 58
Otus scops, 120
Ouzel, Ring, *Turdus torquatus,* 157
Ovenbird, *Seiurus aurocapilla,* 181
Owl, Barn, *Tyto alba,* 121
  Hawk, *Surnia ulula,* 120
  Little, *Athene noctua,* 120
  Long-eared, *Asio otus,* 121
  Scops, *Otus scops,* 120
  Short-eared, *Asio flammeus,* 121
  Snowy, *Bubo scandiacus,* 121
  Tawny, *Strix aluco,* 121
  Tengmalm's, *Aegolius funereus,* 120
Oxyura jamaicensis, 44
Oystercatcher, *Haematopus ostralegus,* 67

Pagophila eburnea, 90
Pandion haliaetus, 115
Panurus biarmicus, 133
Parakeet, Ringed-necked, *Psittacula krameri,* 130
Parkesia noveboracensis, 181

Partridge, Grey, *Perdix perdix,* 30
  Red-legged, *Alectoris rufa,* 30
Parula, Northern, *Setophaga americana,* 140
Parus major, 133
Passer domesticus, 166
  hispaniolensis, 166
  montanus, 166
Passerculus sandwichensis, 180
Passerina cyanea, 174
Pastor roseus, 154
Pelican, Dalmatian, *Pelecanus crispus,* 109
Pelecanus crispus, 109
Perdix perdix, 30
Peregrine, *Falco peregrinus,* 126
Periparus ater, 132
Pernis apivorus, 116
Petrel, Black-capped, *Pterodroma hasitata,* 106
  European Storm, *Hydrobates pelagicus,* 105
  Leach's Storm, *Hydrobates leucorhous,* 105
  Madeiran, *Hydrobates castro,* 182
  Soft-plumaged, *Pterodroma mollis,* 182
  Swinhoe's Storm, *Hydrobates monorhis,* 105
  White-chinned, *Procellaria aequinoctialis,* 106
  Zino's, *Pterodroma madeira,* 106
Petrochelidon pyrrhonota, 137
Phaethon aethereus, 96
Phalacrocorax carbo, 110
Phalarope,
  Grey, *Phalaropus fulicarius,* 81
  Red-necked, *Phalaropus lobatus,* 81
  Wilson's, *Phalaropus tricolor,* 81
Phalaropus fulicarius, 81
  lobatus, 81
  tricolor, 81
Phasianus colchicus, 31
Pheasant, (Common), *Phasianus colchicus,* 31
  Golden, *Chrysolophus pictus,* 31
  Lady Amherst's, *Chrysolophus amherstiae,* 31
Pheucticus ludovicianus, 174

Phoebe, Eastern, *Sayornis phoebe,* 127
Phoenicurus moussieri, 162
   ochruros, 162
   phoenicurus, 162
Phylloscopus bonelli, 141
   borealis, 143
   collybita, 141
   collybita tristis, 141
   coronatus, 143
   fuscatus, 142
   griseolus, 142
   humei, 142
   ibericus, 141
   inornatus, 142
   nitidus, 143
   orientalis, 143
   plumbeitarsus, 143
   proregulus, 142
   schwarzi, 142
   sibilatrix, 141
   tenellipes, 143
   trochiloides, 143
   trochilus, 141
Pica pica, 130
Picus viridis, 123
Pinicola enucleator, 173
Pintail, *Anas acuta,* 41
Pipilo erythrophthalmus, 181
Pipit, Blyth's, *Anthus godlewskii,* 168
   Buff-bellied, *Anthus rubescens,* 169
   Meadow, *Anthus pratensis,* 169
   Olive-backed, *Anthus hodgsoni,* 169
   Pechora, *Anthus gustavi,* 169
   Red-throated, *Anthus cervinus,* 169
   Richard's, *Anthus richardi,* 168
   Rock, *Anthus petrosus,* 168
   Tawny, *Anthus campestris,* 168
   Tree, *Anthus trivialis,* 169
   Water, *Anthus spinoletta,* 168
Piranga olivacea, 174
   rubra, 174
Platalea leucorodia, 109
Plectrophenax nivali, 176
Plegadis falcinellus, 110
Plover, American Golden, *Pluvialis dominica,* 69
   Caspian, *Anarhynchus asiaticus,* 71
   Greater Sand, *Anarhynchus leschenaultii,* 71
   Grey, *Pluvialis squatarola,* 69
   Kentish, *Anarhynchus alexandrinus,* 71
   Lesser Sand, *Anarhynchus mongolus,* 71
   Little Ringed, *Charadrius dubius,* 70
   Pacific Golden, *Pluvialis fulva,* 69
   Ringed, *Charadrius hiaticula,* 70
   Semipalmated, *Charadrius semipalmatus,* 70
Pluvialis dominica, 69
   fulva, 69
   squatarola, 69
Pochard,
   Common, *Aythya ferina,* 44
   Red-crested, *Netta rufina,* 44
Podiceps, auritus, 66
   cristatus, 66
   grisegena, 66
   nigricollis, 66
Podilymbus podiceps, 65
Poecile montanus, 132
   palustris, 132
Polysticta stelleri, 47
Porphyrio alleni, 62
   martinica, 62
   porphyrio, 62
Porzana carolina, 63
   porzana, 63
Pratincole, Black-winged, *Glareola nordmanni,* 85
   Collared, *Glareola pratincola,* 85
Procellaria aequinoctialis, 106
Progne subis, 136
Prunella collaris, 166
   modularis, 166
   montanella, 166
Psittacula krameri, 130
Ptarmigan, *Lagopus muta,* 29
Pterodroma hasitata, 106
   madeira, 106
   mollis, 182
Ptyonoprogne rupestris, 136
Puffin,
   Atlantic, *Fratercula arctica,* 102

Tufted, *Fratercula cirrhata,* 102
Puffinus baroli, 108
  mauretanicus, 108
  puffinus, 108
  yelkouan, 108
Pyrrhocorax pyrrhocorax, 131
Pyrrhula pyrrhula, 170

Quail, *Coturnix coturnix,* 30

Rail, Sora, *Porzana carolina,* 63
  Water, *Rallus aquaticus,* 62
Rallus aquaticus, 62
Raven, *Corvus corax,* 131
Razorbill, *Alca torda,* 101
Recurvirostra avosetta, 63
Redhead, *Aythya americana,* 44
Redpoll, *Acanthis flammea,* 172
Redshank, (Common),
      Tringa totanus, 82
  Spotted, *Tringa erythropus,* 82
Redstart, American, *Setophaga
      ruticilla,* 139
  Black, *Phoenicurus ochruros,* 162
  (Common), *Phoenicurus
      phoenicurus,* 162
  Moussier's, *Phoenicurus
      moussieri,* 162
Redwing, *Turdus iliacus,* 156
Regulus ignicapilla, 151
  regulus, 151
Remiz pendulinus, 133
Rhodostethia rosea, 86
Ring-necked Duck, *Aythya
      collaris,* 45
Riparia riparia, 136
Rissa tridactyla, 86
Robin, American, *Turdus
      migratorius,* 157
  Rufous Bush, *Cercotrichas
      galactotes,* 159
  Rufous-tailed, *Larvivora
      sibilans,* 159
  Siberian Blue, *Larvivora
      cyane,* 160
  White-throated, *Irania
      gutturalis,* 160
Roller, *Coracias garrulus,* 122
Rook, *Corvus frugilegus,* 131
Rosefinch, Common, *Carpodacus
      erythrinus,* 173
Rubythroat, Siberian, *Calliope
      calliope,* 160
Ruff, *Calidris pugnax,* 75

Sanderling, *Calidris alba,* 77
Sandgrouse, Pallas's, *Syrrhaptes
      paradoxus,* 61
Sandpiper,
  Baird's, *Calidris bairdii,* 77
  Broad-billed, *Calidris
      falcinellus,* 77
  Buff-breasted, *Calidris
      subruficollis,* 75
  Common, *Actitis hypoleucos,* 83
  Curlew, *Calidris ferruginea,* 74
  Green, *Tringa ochropus,* 83
  Least, *Calidris minutilla,* 76
  Marsh, *Tringa stagnatilis,* 84
  Pectoral, *Calidris melanotos,* 75
  Purple, *Calidris maritima,* 78
  Semipalmated, *Calidris pusilla,* 78
  Sharp-tailed, *Calidris
      acuminata,* 75
  Solitary, *Tringa solitaria,* 83
  Spotted, *Actitis macularius,* 83
  Stilt, *Calidris himantopus,* 74
  Terek, *Xenus cinereus,* 84
  Upland, *Bartramia longicauda,* 72
  Western, *Calidris mauri,* 78
  White-rumped, *Calidris
      fuscicollis,* 77
  Wood, *Tringa glareola,* 84
Sapsucker, Yellow-bellied,
      Sphyrapicus varius, 123
Saxicola rubetra, 163
  maurus, 163
  rubicola, 163
  stejnegeri, 163
  variegatus, 163
Sayornis phoebe, 127
Seiurus aurocapilla, 181
Serin, *Serinus serinus,* 171
Serinus serinus, 171
Setophaga aestiva, 140
  americana, 140
  castanea, 140
  citrina, 139
  coronata, 138
  fusca, 140

  magnolia, 140
  pensylvanica, 138
  ruticilla, 139
  striata, 138
  tigrina, 140
Scaup, (Greater), *Aythya marila,* 45
 Lesser, *Aythya affinis,* 45
Scoter, Black, *Melanitta americana,* 49
 Common, *Melanitta nigra,* 49
 Stejneger's, *Melanitta stejnegeri,* 49
 Surf, *Melanitta perspicillata,* 49
 Velvet, *Melanitta fusca,* 49
 White-winged, *Melanitta deglandi,* 49
Shag, *Gulosus aristotelis,* 110
Shearwater, Balearic, *Puffinus mauretanicus,* 108
 Barolo, *Puffinus baroli,* 108
 Cory's, *Calonectris borealis,* 107
 Great, *Ardenna gravis,* 107
 Manx, *Puffinus puffinus,* 108
 Scopoli's, *Calonectris diomedea,* 107
 Sooty, *Ardenna grisea,* 107
 Yelkouan, *Puffinus yelkouan,* 108
Shelduck, *Tadorna tadorna,* 37
 Ruddy, *Tadorna ferruginea,* 37
Shoveler, *Spatula clypeata,* 42
Shrike, Brown, *Lanius cristatus,* 128
 Daurian, *Lanius isabellinus,* 128
 Great Grey, *Lanius excubitor,* 129
 Lesser Grey, *Lanius minor,* 129
 Long-tailed, *Lanius schach,* 128
 Masked, *Lanius nubicus,* 129
 Red-backed, *Lanius collurio,* 128
 Turkestan, *Lanius phoenicuroides,* 128
 Woodchat, *Lanius senator,* 129
Sibirionetta, formosa, 37
Siskin, *Spinus spinus,* 171
Sitta canadensis, 152
 europaea, 152
Skua, Arctic, *Stercorarius parasiticus,* 100
 Great, *Stercorarius skua,* 99
 Long-tailed, *Stercorarius longicaudus,* 100
 Pomerine, *Stercorarius pomarinus,* 99
 South Polar, *Stercorarius maccormicki* 99
Skylark, *Alauda arvensis,* 134
Smew, *Mergellus albellus,* 53
Snipe,
 Common, *Gallinago gallinago,* 80
 Great, *Gallinago media,* 80
 Wilson's, *Gallinago delicata,* 80
Somateria, mollissima, 47
 spectabilis, 47
Sparrow, House, *Passer domesticus,* 166
 Lark, *Chondestes grammacus,* 180
 Rock, *Petronia petronia,* 166
 Savannah, *Passerculus sandwichensis,* 180
 Song, *Melospiza melodia,* 180
 Spanish, *Passer hispaniolensis,* 166
 Tree, *Passer montanus,* 166
 White-crowned, *Zonotrichia leucophrys,* 180
 White-throated, *Zonotrichia albicollis,* 180
Sparrowhawk, *Accipiter nisus,* 117
Spatula, clypeata, 42
 querquedula, 38
 discors, 38
Sphyrapicus varius, 123
Spinus spinus, 171
Spoonbill, *Platalea leucorodia,* 109
Starling, (Common), *Sturnus vulgaris,* 154
 Rose-coloured, *Pastor roseus,* 154
Stercorarius longicaudus, 100
 maccormicki, 99
 parasiticus, 100
 pomarinus, 99
 skua, 99
Sterna dougallii, 94
 forsteri, 94
 hirundo, 94
 paradisaea, 94
Sternula albifrons, 93
 antillarum, 93
Stonechat,
 Amur, *Saxicola stejnegeri,* 163
 Caspian, *Saxicola variegatus,* 163

(European), *Saxicola rubicola,* 163
   Siberian, *Saxicola maurus,* 163
Strix aluco, 121
Stilt, Black-winged, *Himantopus himantopus,* 67
Stint, Little, *Calidris minuta,* 76
   Long-toed, *Calidris subminuta,* 76
   Red-necked, *Calidris ruficollis,* 76
   Temminck's, *Calidris temminckii,* 76
Stone-curlew, *Burhinus oedicnemus,* 67
Stork, Black, *Ciconia nigra,* 109
   Whte, *Ciconia ciconia,* 109
Streptopelia decaocto, 61
   orientalis, 61
   turtur, 60
Sturnus vulgaris, 154
Sula leucogaster, 98
   sula, 98
Surnia ulula, 120
Swallow, (Barn), *Hirundo rustica,* 137
   Cliff, *Petrochelidon pyrrhonota,* 137
   Eastern Red-rumped, *Cecropis daurica,* 182
   Red-rumped, *Cecropis rufula,* 137
   Tree, *Tachycineta bicolor,* 136
Swamphen, Western, *Porphyrio porphyrio,* 62
Swan, Bewick's, *Cygnus columbianus,* 36
   Mute, *Cygnus olor,* 36
   Whooper, *Cygnus cygnus,* 36
Swift,
   Alpine, *Tachymarptis melba,* 56
   Chimney, *Chaetura pelagica,* 57
   (Common), *Apus apus,* 56
   Little, *Apus affinis,* 57
   Pacific, *Apus pacificus,* 57
   Pallid, *Apus pallidus,* 56
   White-rumped, *Apus caffer,* 57
Sylvia atricapilla, 148
   borin, 148
Synthliboramphus antiquus, 102
Syrrhaptes paradoxus, 61

Tachybaptus ruficollis, 65
Tachycineta bicolor, 136
Tachymarptis melba, 56

Tadorna, tadorna, 37
   ferruginea, 37
Tanager, Scarlet, *Piranga olivacea,* 174
   Summer, *Piranga rubra,* 174
Tarsiger cyanurus, 160
Tattler, Grey-tailed, *Tringa brevipes,* 79
Tetrao urogallus, 29
Tetrax tetrax, 58
Teal, (Eurasian), *Anas crecca,* 38
   Baikal, *Spatula formosa,* 38
   Blue-winged, *Spatula discors,* 38
   Green-winged, *Anas carolinensis,* 38
Tern, Aleutian, *Onychoprion aleuticus,* 96
   Arctic, *Sterna paradisaea,* 94
   Black, *Chlidonias niger,* 97
   Bridled, *Onychoprion anaethetus,* 96
   Cabot's, *Thalasseus acuflavidus,* 93
   Caspian, *Hydroprogne caspia,* 95
   Common, *Sterna hirundo,* 94
   Elegant, *Thalasseus elegans,* 95
   Forster's, *Sterna forsteri,* 94
   Gull-billed, *Gelochelidon nilotica,* 93
   Least, *Sternula antillarum,* 93
   Lesser Crested, *Thalasseus bengalensis,* 95
   Little, *Sternula albifrons,* 93
   Roseate, *Sterna dougallii,* 94
   Royal, *Thalasseus maximus,* 95
   Sandwich, *Thalasseus sandvicensis,* 93
   Sooty, *Onychoprion fuscatus,* 96
   Whiskered, *Chlidonias hybrida,* 97
   White-winged, *Chlidonias leucopterus,* 97
Thalassarche chlororhynchos, 104
   melanophris, 104
Thalasseus acuflavidus, 93
   bengalensis, 95
   elegans, 95
   maximus, 95
   sandvicensis, 95
Thrasher, Brown, *Toxostoma rufum,* 153

Thrush, Black-throated, *Turdus atrogularis,* 158
  Blue Rock, *Monticola solitarius,* 162
  Dusky, *Turdus eunomus,* 158
  Eyebrowed, *Turdus obscurus,* 156
  Grey-cheeked, *Catharus minimus,* 155
  Hermit, *Catharus guttatus,* 155
  Mistle, *Turdus viscivorus,* 156
  Naumann's, *Turdus naumanni,* 158
  Red-throated, *Turdus ruficollis,* 158
  Rock, *Monticola saxatilis,* 162
  Siberian, *Geokichla sibirica,* 158
  Song, *Turdus philomelos,* 156
  Swainson's, *Catharus ustulatus,* 155
  Varied, *Ixoreus naevius,* 155
  White's, *Zoothera aurea,* 156
  Wood, *Hylocichla mustelina,* 155
Tichodroma muraria, 152
Tit, Bearded, *Panurus biarmicus,* 133
  Blue, *Cyanistes caeruleus,* 133
  Coal, *Periparus ater,* 132
  Crested, *Lophophanes cristatus,* 132
  Great, *Parus major,* 133
  Long-tailed, *Aegithalos caudatus,* 132
  Marsh, *Poecile palustris,* 132
  Penduline, *Remiz pendulinus,* 133
  Willow, *Poecile montanus,* 132
Towhee, Eastern, *Pipilo erythrophthalmus,* 181
Toxostoma rufum, 153
Treecreeper, (Eurasian), *Certhia familiaris,* 152
  Short-toed, *Certhia brachydactyla,* 152
Tringa brevipes, 79
 erythropus, 82
 flavipes, 82
 glareola, 84
 melanoleuca, 82
 ochropus, 83
 solitaria, 83
 stagnatilis, 84
 totanus, 82
Troglodytes troglodytes, 152

Tropicbird, Red-billed, *Phaethon aethereus,* 96
Turdus atrogularis, 158
 eunomus, 158
 iliacus, 156
 merula, 157
 migratorius, 157
 naumanni, 158
 obscurus, 156
 philomelos, 156
 pilaris, 157
 ruficollis, 158
 torquatus, 157
 viscivorus, 156
Turnstone, *Arenaria interpres,* 79
Twite, *Linaria flavirostris,* 172
Tyrannus tyrannus, 127
Tyto alba, 121

Upupa epops, 122
Uria aalge, 101
 lomvia, 101

Vanellus cinereus, 82
 gregarius, 68
 leucurus, 68
 vanellus, 68
Veery, *Catharus fuscescens,* 155
Vermivora chrysoptera, 139
Vireo, Philadelphia, *Vireo philadelphicus,* 138
  Red-eyed, *Vireo olivaceus,* 138
  Yellow-throated, *Vireo flavifrons,* 138
Vireo flavifrons, 138
 olivaceus, 138
 philadelphicus, 138
Vulture, Egyptian, *Neophron percnopterus,* 115

Wagtail, Citrine, *Motacilla citreola,* 167
  Eastern Yellow, *Motacilla tschutschensis,* 167
  Grey, *Motacilla cinerea,* 167
  Pied, *Motacilla alba yarrellii,* 167
  White, *Motacilla alba,* 167
Wallcreeper, *Tichodroma muraria,* 152
Waterthrush, Northern, *Parkesia*

noveboracensis, 181
Warbler, Aquatic, *Acrocephalus paludicola,* 144
Arctic, *Phylloscopus borealis,* 143
Asian Desert, *Curruca nana,* 150
Barred, *Curruca nisoria,* 148
Bay-breasted, *Setophaga castanea,* 140
Black-and-white, *Mniotilta varia,* 140
Blackburnian, *Setophaga fusca,* 140
Blackpoll, *Setophaga striata,* 138
Blyth's Reed, *Acrocephalus dumetorum,* 145
Booted, *Iduna caligata,* 146
Canada, *Cardellina canadensis,* 139
Cape May, *Setophaga tigrina,* 140
Cetti's, *Cettia cetti,* 145
Chestnut-sided, *Setophaga pensylvanica,* 138
Dartford, *Curruca undata,* 150
Dusky, *Phylloscopus fuscatus,* 142
Eastern Bonelli's, *Phylloscopus orientalis,* 141
Eastern Crowned, *Phylloscopus coronatus,* 143
Eastern Olivaceous, *Iduna pallida,* 146
Eastern Orphean, *Curruca crassirostris,* 148
Eastern Subalpine, *Curruca cantillans,* 149
Fan-tailed, *Cisticola juncidis,* 144
Garden, *Sylvia borin,* 148
Golden-winged, *Vermivora chrysoptera,* 139
Grasshopper, *Locustella naevia,* 147
Great Reed, *Acrocephalus arundinaceus,* 144
Green, *Phylloscopus nitidus,* 143
Greenish, *Phylloscopus trochiloides,* 143
Hooded, *Setophaga citrina,* 139
Hume's, *Phylloscopus humei,* 142
Icterine, *Hippolais icterina,* 146
Lanceolated, *Locustella lanceolata,* 147
Magnolia, *Setophaga magnolia,* 140
Marmora's, *Curruca sarda,* 150
Marsh, *Acrocephalus palustris,* 145
Melodious, *Hippolais polyglotta,* 146
Moltoni's, *Curruca subalpina,* 149
Myrtle, *Setophaga coronata,* 138
Olive tree, *Hippolais olivetorum,* 146
Paddyfield, *Acrocephalus agricola,* 145
Pale-legged Leaf, *Phylloscopus tenellipes,* 143
Pallas's, *Phylloscopus proregulus,* 142
Pallas's Grasshopper, *Helopsaltes certhiola,* 147
Radde's, *Phylloscopus schwarzi,* 142
Reed, *Acrocephalus scirpaceus,* 145
River, *Locustella fluviatilis,* 147
Rüppell's, *Curruca ruppeli,* 149
Sardinian, *Curruca melanocephala,* 149
Savi's, *Locustella luscinioides,* 147
Sedge, *Acrocephalus schoenobaenus*, 144
Spectacled, *Curruca conspicillata,* 150
Sulphur-bellied, *Phylloscopus griseolus,* 142
Sykes's, *Iduna rama,* 146
Tennessee, *Leiothlypis peregrina,* 139
Thick-billed, *Arundinax aedon,* 144
Two-barred Greenish, *Phylloscopus plumbeitarsus,* 143
Western Bonelli's, *Phylloscopus bonelli,* 141
Western Olivaceous, *Iduna opaca,* 146
Western Orphean, *Curruca hortensis,* 148
Western Subalpine, *Curruca iberiae,* 149
Willow, *Phylloscopus*

trochilus, 141
Wilson's, *Cardellina pusilla,* 138
Wood, *Phylloscopus sibilatrix,* 141
Yellow, *Setophaga aestiva,* 140
Yellow-browed, *Phylloscopus inornatus,* 142
Waxwing, (Bohemian), *Bombycilla garrulus,* 153
Cedar, *Bombycilla cedrorum,* 153
Wheatear, Black-eared, *Oenanthe hispanica,* 165
Western Black-eared *Oenanthe hispanica hispanica,* 165
Eastern Black-eared, *Oenanthe hispanica melanoleuca* 165
Desert, *Oenanthe deserti,* 164
Isabelline, *Oenanthe isabellina,* 164
(Northern), *Oenanthe oenanthe,* 164
Pied, *Oenanthe pleschanka,* 165
White-crowned, *Oenanthe leucopyga,* 165
Whimbrel, (Eurasian), *Numenius phaeopus,* 72
Hudsonian, *Numenius hudsonicus,* 72
Little, *Numenius minutus,* 72
Whinchat, *Saxicola rubetra,* 163
Whitethroat, (Common), *Curruca communis,* 150
Lesser, *Curruca curruca,* 148
Siberian Lesser, *Curruca curruca blythi,* 148
Wigeon, (Eurasian), *Mareca penelope,* 42
American, *Mareca americana,* 42
Woodlark, *Lullula arborea,* 134
Woodpecker, Great Spotted, *Dendrocopos major,* 123
Green, *Picus viridis,* 123
Lesser-spotted, *Dryobates minor,* 123
Woodpigeon, *Columba palumbus,* 60
Wren, *Troglodytes troglodytes,* 152
Wryneck, *Jynx torquilla,* 123

Xema sabini, 86
Xenus cinereus, 84

Yellowhammer, *Emberiza citrinella,* 178
Yellowlegs, Greater, *Tringa melanoleuca,* 82
Lesser, *Tringa flavipes,* 82
Yellowthroat, Common, *Geothlypis trichas,* 139

Zapornia parva, 63
pusilla, 63
Zenaida macroura, 60
Zonotrichia albicollis, 180
leucophrys, 180
Zoothera aurea, 156

Parts of a bird

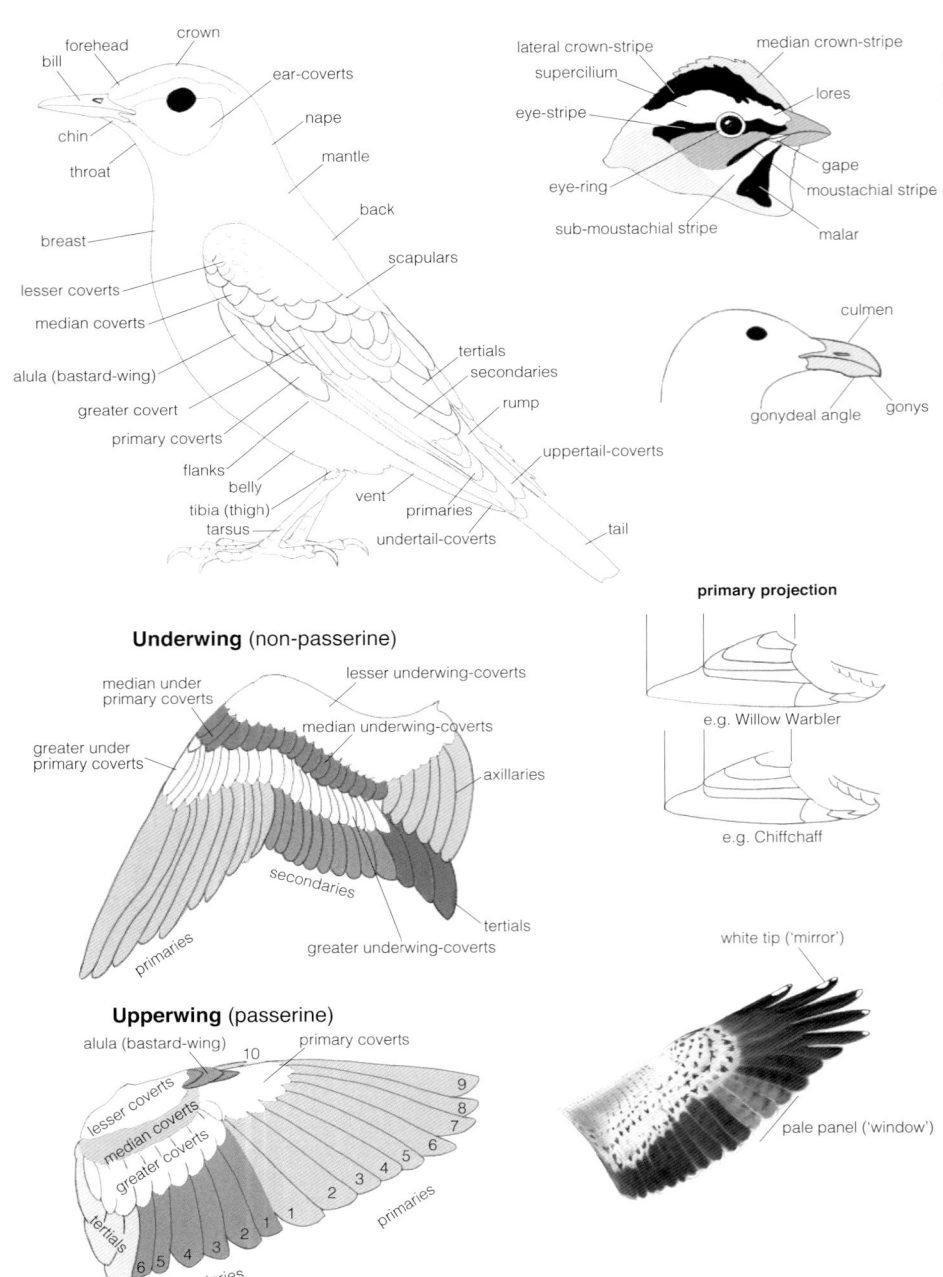